Advances in Experimental Medicine and Biology

Volume 1083

Innovations in Cancer Research and Regenerative Medicine

This book series represents a major contribution to the emerging science of cancer research and regenerative medicine. The volumes bring together some of the most pre-eminent scientists working on cancer biology, cancer treatment, cancer diagnosis, cancer prevention and regenerative medicine to share information on currently ongoing work that will help shape future therapies.

The book series publishes review and original research contributions, short reports, as well as guest-edited thematic book volumes. All contributions will be published online first and collected in book volumes. There are no publication costs.

Innovations in Cancer Research and Regenerative Medicine is a subseries of *Advances in Experimental Medicine and Biology*, which has been publishing significant contributions in the field for over 30 years and is indexed in Medline, Scopus, EMBASE, BIOSIS, Biological Abstracts, CSA, Biological Sciences and Living Resources (ASFA-1), and Biological Sciences. The 2016 Impact Factor is 1.881.

More information about this subseries at https://www.springer.com/series/15740

Phuc Van Pham
Editor

Stem Cells: Biology and Engineering

Springer

Editor
Phuc Van Pham
Laboratory of Stem Cell Research and Application
University of Science, Vietnam National University
Ho Chi Minh City, Vietnam

ISSN 0065-2598 ISSN 2214-8019 (electronic)
Advances in Experimental Medicine and Biology
ISSN 2524-6100 ISSN 2524-6119 (electronic)
Innovations in Cancer Research and Regenerative Medicine
ISBN 978-3-319-77481-7 ISBN 978-3-319-77482-4 (eBook)
https://doi.org/10.1007/978-3-319-77482-4

Library of Congress Control Number: 2018945065

This Springer imprint is published by the registered company Springer Nature Switzerland AG
The registered company address is: Gewerbestrasse 11, 6330 Cham, Switzerland

Contents

Contributors

Ladislaus Adam Department of Immunology and Institute for Cellular and Molecular Medicine; SAMRC Extramural Unit for Stem Cell Research and Therapy; Faculty of Health Sciences, University of Pretoria, Pretoria, South Africa

Melvin A. Ambele Department of Immunology and Institute for Cellular and Molecular Medicine, SAMRC Extramural Unit for Stem Cell Research and Therapy, Faculty of Health Sciences, University of Pretoria, Pretoria, South Africa

Department of Oral Pathology and Oral Biology, School of Dentistry, Faculty of Health Sciences, University of Pretoria, Pretoria, South Africa

Hossein Azizi Faculty of Biotechnology, Amol University of Special Modern Technologies, Amol, Iran

Mohamad Al Bazzal Lebanese University School of Dentistry, Beirut, Lebanon

Sonja C. Boy Department of Oral Pathology, School of Oral Health Sciences, Sefako Makgatho Health Sciences University, Ga-Rankuwa, Pretoria, South Africa

Anh Nguyen-Tu Bui Stem Cell Institute, University of Science VNUHCM, Ho Chi Minh City, Vietnam

Anh Thi-Van Bui Stem Cell Institute, University of Science VNUHCM, Ho Chi Minh City, Vietnam

Laboratory of Stem Cell Research and Application, University of Science VNUHCM, Ho Chi Minh City, Vietnam

Sabine Conrad Tübingen, Germany

Loan Thi-Tung Dang Stem Cell Institute, University of Science VNUHCM, Ho Chi Minh City, Vietnam

Thuy Thi-Thanh Dao Laboratory of Stem Cell Research and Application, University of Science, Vietnam National University, Ho Chi Minh City, Vietnam

Chrisna Durandt Department of Immunology and Institute for Cellular and Molecular Medicine; SAMRC Extramural Unit for Stem Cell Research and Therapy; Faculty of Health Sciences, University of Pretoria, Pretoria, South Africa

Nazmul Haque Department of Restorative Dentistry, Faculty of Dentistry, University of Malaya, Kuala Lumpur, Malaysia

Regenerative Dentistry Research Group, Faculty of Dentistry, University of Malaya, Kuala Lumpur, Malaysia

Rasime Kalkan Faculty of Medicine, Department of Medical Genetics, Near East University, Nicosia, Cyprus

Noor Hayaty Abu Kasim Department of Restorative Dentistry, Faculty of Dentistry, University of Malaya, Kuala Lumpur, Malaysia

Regenerative Dentistry Research Group, Faculty of Dentistry, University of Malaya, Kuala Lumpur, Malaysia

Ngoc Phan Kim Stem Cell Institute, University of Science VNUHCM, Ho Chi Minh City, Vietnam

Ha Thi-Ngan Le Laboratory of Stem Cell Research and Application, University of Science, Vietnam National University, Ho Chi Minh City, Vietnam

Anne E. Mercier Department of Physiology, Faculty of Health Sciences, University of Pretoria, Pretoria, South Africa

Ibrahim Mortada Faculty of Medicine, American University of Beirut, Beirut, Lebanon

Rola Mortada Lebanese University School of Dentistry, Beirut, Lebanon

Sulette Nel Department of Oral Pathology and Oral Biology, School of Dentistry, Faculty of Health Sciences, University of Pretoria, Pretoria, South Africa

Cong Le-Thanh Nguyen Stem Cell Institute, University of Science VNUHCM, Ho Chi Minh City, Vietnam

Laboratory of Stem Cell Research and Application, University of Science VNUHCM, Ho Chi Minh City, Vietnam

Ahmed Kolade Oloyo Institute for Cellular and Molecular Medicine, Department of Immunology, and SAMRC Extramural Unit for Stem Cell Research and Therapy, Faculty of Health Sciences, University of Pretoria, Pretoria, South Africa

Department of Physiology, Faculty of Basic Medical Science, College of Medicine, University of Lagos, Lagos, Nigeria

Michael Sean Pepper Institute for Cellular and Molecular Medicine (ICMM), Department of Immunology, and SAMRC Extramural Unit for Stem Cell Research and Therapy, Faculty of Health Sciences, University of Pretoria, Pretoria, South Africa

Phuc Van Pham Stem Cell Institute, University of Science, VNUHCM, Ho Chi Minh City, Vietnam

Laboratory of Stem Cell Research and Application, University of Science, VNUHCM, Ho Chi Minh city, Vietnam

Ngoc Kim Phan Laboratory of Stem Cell Research and Application, University of Science, Vietnam National University, Ho Chi Minh City, Vietnam

Lan Thi Phi Laboratory of Stem Cell Research and Application, University of Science, Vietnam National University, Ho Chi Minh City, Vietnam

Marnie Potgieter Department of Immunology and Institute for Cellular and Molecular Medicine; SAMRC Extramural Unit for Stem Cell Research and Therapy; Faculty of Health Sciences, University of Pretoria, Pretoria, South Africa

Center for Microbial Ecology and Genomics, Department of Genetics, Faculty of Natural and Agricultural Sciences, University of Pretoria, Pretoria, South Africa

Nedime Serakinci Faculty of Medicine, Department of Medical Genetics, Near East University, Nicosia, Cyprus

Faculty of Arts and Sciences, Department of Molecular Biology and Genetics, Near East University, Nicosia, Cyprus

Thomas Skutella Institute for Anatomy and Cell Biology III, Medical Faculty, Heidelberg University, Heidelberg, Germany

Van Thanh Ta Department of Biochemistry, Hanoi Medical University, Hanoi, Vietnam

Kiet Dinh Truong Medical Genetics Institute, Ho Chi Minh City, Vietnam

Nhat Chau Truong Stem Cell Institute, University of Science VNUHCM, Ho Chi Minh City, Vietnam

Laboratory of Stem Cell Research and Application, University of Science VNUHCM, Ho Chi Minh City, Vietnam

Pinar Tulay Faculty of Medicine, Department of Medical Genetics, Near East University, Nicosia, Cyprus

Danielle de Villiers Department of Immunology and Institute for Cellular and Molecular Medicine; SAMRC Extramural Unit for Stem Cell Research and Therapy; Faculty of Health Sciences, University of Pretoria, Pretoria, South Africa

Ngoc Bich Vu Laboratory of Stem Cell Research and Application, University of Science, Vietnam National University, Ho Chi Minh City, Vietnam

Elize Wolmarans Institute for Cellular and Molecular Medicine (ICMM), Department of Immunology, and SAMRC Extramural Unit for Stem Cell Research and Therapy, Faculty of Health Sciences, University of Pretoria, Pretoria, South Africa

Qing-Shuo Zhang Oregon Stem Cell Center, Department of Pediatrics, Oregon Health & Science University, Portland, OR, USA

Advs Exp. Medicine, Biology - Innovations in Cancer Research and Regenerative Medicine (2018) 1: 1–17
DOI 10.1007/5584_2017_63

Published online: 8 July 2017

Allogeneic Adipose-Derived Mesenchymal Stem Cell Transplantation Enhances the Expression of Angiogenic Factors in a Mouse Acute Hindlimb Ischemic Model

Ngoc Bich Vu, Ha Thi-Ngan Le, Thuy Thi-Thanh Dao, Lan Thi Phi, Ngoc Kim Phan, and Van Thanh Ta

Abstract

Cell migration and molecular mechanisms during healing of damaged vascular or muscle tissues are emerging fields of interest worldwide. The study herein focuses on evaluating the role of allogenic adipose-derived mesenchymal stem cells (ADMSCs) in restoring damaged tissues. Using a hindlimb ischemic mouse model, ADMSC-mediated induction of cell migration and gene expression related to myocyte regeneration and angiogenesis were evaluated. ADMSCs were labeled with GFP (ADMSC-GFP). The proximal end of the femoral blood vessel of mice (over 6 months of age) are ligated at two positions then cut between the two ties. Hindlimb ischemic mice were randomly divided into two groups: Group I ($n = 30$) which was injected with PBS (100 μL) and Group II ($n = 30$) which was transplanted with ADMSC-GFP (10^6 cells/100 μL PBS) at the rectus femoris muscle. The migration of ADMSC-GFP in hindlimb was analyzed by UV-Vis system. The expression of genes related to angiogenesis and muscle tissue repair was quantified by real-time RT-PCR. The results showed that ADMSCs existed in the grafted hindlimb for 7 days. Grafted cells migrated to other damaged areas such as thigh and heel. In both groups the ischemic hindlimb showed an increased expression of several angiogenic genes, including *Flt-1*, *Flk-1*, and *Ang-2*. In particular, the expression of Ang-2 and myogenic-related gene MyoD was significantly increased in the ADMSC-treated group compared to the

N.B. Vu (✉), H.T.-N. Le, T.T.-T. Dao, L.T. Phi, and N.K. Phan
Laboratory of Stem Cell Research and Application, University of Science, Vietnam National University, 6 Quarter, Linh Trung Ward, Thu Duc District, Ho Chi Minh City, Vietnam
e-mail: vbngoc@hcmus.edu.vn; ltnha@hcmus.edu.vn; dttthuy@hcmus.edu.vn; ptlan@hcmus.edu.vn; pkngoc@hcmus.edu.vn

V.T. Ta
Department of Biochemistry, Hanoi Medical University, No 1, Ton That Tung St., Dong Da, Hanoi, Vietnam
e-mail: tathanhvan@hmu.edu.vn

PBS-treated (control) group; the expression increased at day 28 compared to day 3. The other factors, such as *VE-Cadherin*, *HGF*, *CD31*, *Myf5*, and *TGF-β*, were also more highly expressed in the ADMSC-treated group than in the control group. Thus, grafted ADMSCs were able to migrate to other areas in the injured hindlimb, persist for approximately 7 days, and have a significantly positive impact on stimulating expression of myogenic- and angiogenesis-related genes.

Keywords
Adipose tissue • Angiogenesis • Hindlimb ischemia • Mesenchymal stem cell • Myogenesis

Abbreviations

ADMSC	Adipose-derived mesenchymal stem cells
bFGF	Basic fibroblast growth factor
CD	Cluster of differentiation
EGF	Epidermal growth factor
GFP	Green fluorescent protein
HGF	Hepatic growth factor
MSC	Mesenchymal stem cell
PBS	Phosphate buffer saline
VEGF	Vascular endothelial growth factor

1 Introduction

The potential of mesenchymal stem cells (MSCs), especially adipose-derived stem cells, to treat a variety of diseases has bolstered interest and popularized the study of their functions and applications. Due to their ability to differentiate into osteoblasts, chondrocytes, adipocytes, and cardiomyocytes, MSCs represent an exciting prospective cell source for the treatment of many diseases (Pipino and Pandolfi 2015; Strem et al. 2005; Toma et al. 2002; Zhuang et al. 2016). Moreover, MSCs are appropriate for applications in transplantation since they have a potent ability to regulate immune responses (Kim and Cho 2015).

Researchers have recognized the exclusive benefits of MSCs in healing and regenerating tissue. Firstly, when MSCs are transplanted in the damaged site, they can secrete cytokines and growth factors to stimulate tissue regeneration; this is called the paracrine pathway (Baraniak and McDevitt 2010). On the other hand, MSCs can regulate stem cell niches in the host by mobilizing stem cells to the injured tissue and inducing differentiation into essential cell types. Additionally, MSCs can produce antioxidative factors. This means MSCs can help clean up damaged cells and prevent oxidative stress (Fouraschen et al. 2015). Similarly, MSCs can help repair injured tissues by two mechanisms: (1) stem cells can directly differentiate into functional cells or fuse with damaged cells to replace, repair, or renew the necrotic/damaged cell, and (2) stem cells can secrete several factors in order to attract another specific cell type to join in the regeneration. The secreted factors may be paracrine or autocrine elements and may help protect functional cells from apoptosis. Moreover, they can be factors which activate and mobilize endogenous stem cells to home to areas of tissue injury (Rennert et al. 2012). The complex movement of stem cells is regulated by a vast array of cytokines, adhesion molecules, and essential growth factors (Smart and Riley 2008).

The ischemic model originally developed in mouse was based on a previously developed rabbit model. The hindlimb ischemic model was operatively induced in a mouse by disruption of blood flow to the lower limb; this leads to the dysfunction and/or damage of cells/tissues under the blood vessel cutting site. Research studies on the capability of MSCs to recover ischemic injury, by stimulating angiogenesis in vitro and in vivo, have been conducted (Li et al. 2015; Sun et al. 2015). The number of new vessels after ischemia has been shown to be correlated with the recovery ability of patients with ischemic diseases.

The term "angiogenesis" was first introduced in 1935 by Hertig to describe the formation of new blood vessels in the placenta (Hertig 1935). Angiogenesis is defined as the formation of new blood vessels via the sprouting of endothelial cells from preexisting vasculature (Ucuzian et al. 2010). In adults, angiogenesis is relatively rare (Gerhardt and Betsholtz 2003). This process usually occurs after injury, during pregnancy, and during tumor invasion (Nishida et al. 2006; Zygmunt et al. 2003). For the sprouting of new vessels, endothelial cells undergo a complicated process. From the degradation of the underlying basement membrane, endothelial progenitor cells migrate into the neighboring tissues. At the new site, they proliferate, differentiate, and assemble into tubes. Finally, the tube-to-tube connection is formed for blood flow. Overall, angiogenesis is a process which includes an increase in vascular permeability, degradation of the surrounding matrix, proliferation and migration of endothelial cells, and finally stabilization (Conway et al. 2001). To date, there has been considerable effort invested in understanding the molecular mechanisms of angiogenesis following tissue injury. This is still a complex process which requires the participation of many regulator factors, cytokines, cell/matrix interactions, and pertinent intracellular signaling pathways (Buschmann and Schaper 1999).

Research studies have demonstrated that MSCs express some pre-angiogenic factors (Kamihata et al. 2001). Particularly, adipose-derived mesenchymal stem cells (ADMSCs) can secrete several angiogenic growth factors, including vascular endothelial growth factor (VEGF), hepatocyte growth factor (HGF), and basic fibroblast growth factor (bFGF) (Cui et al. 2015). Soluble growth factors released by ADMSCs play an important role in the proliferation and migration of cells and smooth muscle cells; in vitro studies have shown that this ability is dependent on the number of MSCs (Iwase et al. 2005; Zentilin et al. 2006). VEGF, also known as vascular permeability factor, is a dimeric glycoprotein that acts as a mitogen for endothelial cells (Dvorak et al. 1995). To date, VEGF is the best candidate to act as a specific regulator of endothelial cell growth and differentiation (Marti et al. 2000). The formation of new vessels occurs in parallel with an increase of VEGF mRNA and protein expression. HGF regulates cell growth, cell motility, and morphogenesis of different kinds of cell types (Hayashi et al. 1999; Nakamura et al. 2000). Similar to VEGF, HGF is secreted by intact cells and acts as a mitogen for endothelial cells. However, studies have shown that HGF does not stimulate monocyte migration through endothelial layers like VEGF (Hayashi et al. 1999). While growth factors like VEGF and HGF activate endothelial cells to form new vessels, the PDGFB/PDGFR-beta signaling system and angiopoietins Ang-1 and Ang-2 and their receptor (Tie2) play a role in the stabilization and maturation of vascular wall formation by recruiting pericytes and smooth muscle cells (Karamysheva 2008).

In this study, in a hindlimb ischemic mouse model, we investigated the role of allogeneic ADMSCs in restoring damaged tissue via migration of cells and expression of genes related to muscle cell regeneration and angiogenesis.

2 Methods

2.1 Establishing GFP-Positive ADMSCs

ADMSCs were extracted from adipose tissue by a previously established protocol (Van Pham et al. 2014). ADMSCs were cultured in MSCCult medium (Regenmed Lab, HCM, VN) that DMEM/F12 supplemented with 10% fetal bovine serum, 1% antibiotic, 100x antimycotic, 10 ng/mL EGF, and 10 ng/mL bFGF (Sigma-Aldrich, Louis St, MO) in a humidified incubator with 5% CO_2, at 37 °C. To generate ADMSC-GFP, ADMSCs (passage 3 or beyond) were transferred onto 6-well tissue culture plates. After 24 h, polybrene (8 μg/mL) (Sigma-Aldrich) was added, and 24 h following that, fresh medium (without polybrene) and copGFP Lentiviral Particles (Santa Cruz Biotechnology, Dallas, TX) were added to transduce copGFP into ADMSCs. The cells were cultured in MSCCult medium supplemented with 10 μg/mL puromycin dihydrochloride (Santa Cruz Biotechnology) for a week to select for GFP-positive ADMSCs. After that, culture medium with puromycin was replaced by fresh MSCCult medium. ADMSCs positive for GFP were observed under fluorescence microscope. The proportion of GFP-positive ADMSCs was quantified by FACSCalibur system (BD Biosciences, San Jose, CA). All data were analyzed by CellQuest Pro software at 10,000 events.

2.2 Mouse Model of Hindlimb Ischemia

Adult mice (over 6 months of ages) were used in our study with animal approval in the Stem Cell Laboratory, Research and Application, University of Science, VNU-HCM, Vietnam. The hindlimb ischemic mouse model was established according to our published protocol (Vu et al. Vu et al. 2013). Mice were anesthetized with ketamine (100 mg/kg i.m.) and xylazine (15 mg/kg i.m.). The femoral artery and vein were ligated at two sites and then cut between the two ties, as described in Fig. 1.

2.3 Cell Transplantation

In our acute hindlimb ischemic mouse model, mice were randomly divided into two groups: Group A (ADMSC-GFP treatment, 30 mice) and Group B (PBS treatment, 30 mice). Both groups were injected with the same volume (100 μL), at the same location, and around the same time. The surgical protocol was as follows: mice were anesthetized as mentioned above, then ADMSC-GFP (10^6/100 μL PBS) was prepared for transplantation; ADMSC-GFP were injected to the rectus femoris muscle promptly after the blood vessel was cut; finally, the skin was closed with a nylon suture.

The detection of ADMSC-GFP was observed using the iBox Explorer Imaging Microscope system (UVP, Upland, CA). Images were taken at the site of ADMSC-GFP injection under UV light and analyzed by Vision WorksLS Image Acquisition and Analysis Software.

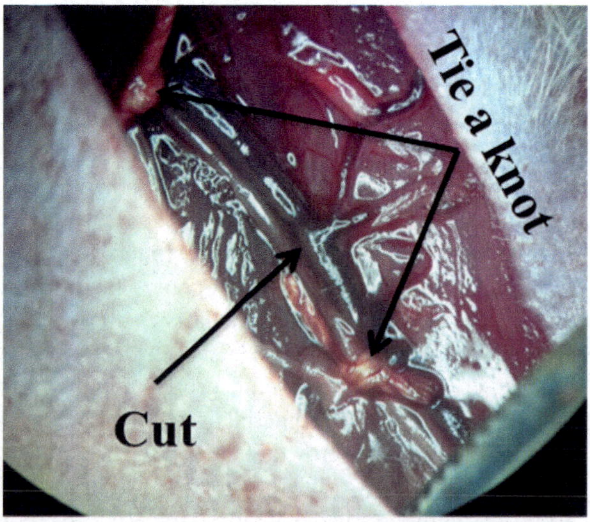

Fig. 1 The blood vessel site was cut to establish the hindlimb ischemic mouse model

2.4 Quantitation of Transplanted Cells in the Host

The percentage of detectable transplanted cells in the ischemic hindlimb was assessed by flow cytometry. Muscle tissue of ADMSC-GFP-transplanted mice was collected from the cell-injected site (IS) and the lateral gastrocnemius site (LGS). The muscle tissues were processed into single cell suspension using 0.5% trypsin/EDTA. The percentage of GFP-positive cells was evaluated by FACSCalibur system (BD Biosciences, San Jose, CA) and analyzed by CellQuest Pro software (BD Biosciences). All data were acquired with an event acquisition set for 10,000 events.

2.5 RNA Extraction

At 1 day, 3 days, and 28 days after ADMSC-GFP transplantation, 100 μg of femoral muscle tissue was collected into 2 mL tubes. Muscle tissue was fine-cut then total RNA was extracted following the manufacturer's protocol using RNA-Spin Total RNA Extraction Kit (Intron Biotechnology, Korea).

Table 1 Primer sequences used in this study

Name of primer	Length (bp)	Primer sequence (5′-3′) (F, forward; R, reverse)	Ref. (NCBI)
GAPDH	80	F: ACTGAGCAAGAGAGGCCCTA	NM_001289726.1
		R: TATGGGGGTCTGGGATGGAA	
PGFRB	156	F:ATCTTGGACCCCACCTGAAC	NM_008809.2
		R:AATGCTTGTTGGAGTGTCGC	
TGF-beta	70	F:CATGGAGCTGGTGAAACGGA	NM_011577.2
		R: GGCGAGCCTTAGTTTGGACA	
MMP-2	123	F: AACGGTCGGGAATACAGCAG	NM_008610.2
		R:AAACAAGGCTTCATGGGGGC	
Ephrin-B2	139	F:ATGTAACAGACGGGGGTTTGA	NM_001290753.2
		R:CCGCACTGCCTCATTAGGAT	
Ephrin-B4	140	F: ATGCCCAGCCAACAGCCACT	NM_001159571.1
		R:AATGGTGAACCACGCTTCTTG	
Flk − 1	200	F: CCTACCCCACACATTACATGG	NM_010612.2
		R: TTTCCTGGGCACCTTCTATT	
Flt-1	186	F: GTGTCTATAGGTGCCGAGCC	NM_010228.3
		R: CGGAAGAAGACCGCTTCAGT	
vWF	71	F: CTACCTAGAACGCGAGGCTG	NM_011708.4
		R: CATCGATTCTGGCCGCAAAG	
VE-cadherin	93	F: CCTGAGGCAATCAACTGTGC	NM_009868.4
		R: GGAGGAGCTGATCTTGTCCG	
CD31	70	F: CTCCCTTGAGCCTCACCAAG	NM_001305157.1
		R: GACCTTCCGGATCTCACTGT	
ANG-1	126	F: ATGGGCAATGTGCCTACACTTT	NM_001286062.1
		R: AAGAGAAATCCGGCTCCACGT	
ANG-2	97	F:AGTACCAGGTCCAGAACGGA	NM_007426.4
		R:ACGGCATTGGACATGTAGGG	
HGF	71	F:CCTGTGCCTTGACTTAGCGA	NM_001289459.1
		R:TCCTTGTGGCTCCTATCCGA	
Myf5	177	F:AGATCCTCAGGAATGCCATCCG	NM_008656.5
		R:TGCTGTTCTTTCGGGACCAGAC	
MyoD	148	F:GCTCTGATGGCATGATGGATTACAGCG	NM_010866.2
		R:ATGCTGGACAGGCAGTCGAGGC	

2.6 Real-Time Quantitative Reverse Transcription PCR

One-step quantitative reverse transcriptase PCR (qRT-PCR) reactions were carried out in an Eppendorf real-time PCR system (Eppendorf, Germany) using 300 ng of total RNA and SYBR Green qRT-PCR Master Mix (PCR Biosystems Ltd., London, UK). The reaction conditions were set for 10 min at 25 °C (reverse transcription), 5 min at 95 °C (polymerase activation), 40 cycles of 15 s at 95 °C (denaturation), 15 s at 58 °C (annealing), and 20 s at 72 °C (extension), with fluorescent signal recording for amplification. At the end, a final step of 15 s at 95 °C and 15 s at 60 °C was included, followed by fluorescence measurement at each 0.5 °C variation from 58 °C to 95 °C in 20 min, to obtain the melting curve. The primer sequences are listed in Table 1.

2.7 Statistics

The data were analyzed by GraphPad Prism 6.0 software, and a t-test was conducted for statistical analysis; P-value less than 0.05 was considered statistically significant.

3 Results

3.1 Transgenic GFP Detection in Adipose Tissue-Derived Mesenchymal Stem Cells

After 24 h of transduction, some cells showed positivity for GFP expression. The number of ADMSC-GFP continued to increase over the next 2 days. After treatment with medium containing 10 μg/mL of puromycin (for selection), GFP-positive cells increased markedly. After transduction, ADMSCs were evaluated for expression of GFP. The percentage of ADMSC-GFP and their morphology were recorded under white light (Fig. 2A) and green fluorescent light (Fig. 2B). The results indicated that 100% of the transduced cells were positive for the GFP transgene when analyzed quantitatively by flow cytometry (Fig. 2C).

3.2 Detection of Allogenic ADMSCs in the Host

The appearance of bright spots was not observed in PBS-injected mice (Fig. 3A). Immediately after transplantation, at the graft position, light spots were expressed strongly (Fig. 3B). The red spot represents a high density of cells. The other ones (yellow and blue spots) represent decreased cell density. After 7 days, the size and density of cells at the sites of grafting were decreased (Fig. 3C). After 10 days, light spots were not observed at the cell-grafted positions.

When observed with the naked eye through the filter of a fluorescence imaging system, grafted cells were clearly detectable by the presentation of green spots (blue arrow, Fig. 3E) compared to the PBS group (Fig. 3D). The movement of transplanted cells was clearly observed after approximately 1–2 h, with the separation of the light spot from one to two spots (Fig. 3F). This suggests that transplanted cells migrated to other locations in the hindlimb of the host.

To evaluate if the stem cells were capable of migrating to the heel and foot positions, which are more vulnerable to shortage of oxygen and nutrients in acute ischemia, we collected whole cells from grafted thigh to detect fluorescence signal along the thigh. The results showed that on the date of the evaluation (day 4 after transplantation), grafted cells were detected (red arrow) and distributed along the length of the thigh and were concentrated at the grafted positions (heel and foot) (Fig. 4).

3.3 Quantitation of Transplanted Cells in the Host

One day after transplantation, the percentage of GFP-positive cells at the IS was 8.12% ± 1.00% (Fig. 5A). This proportion decreased over time, reaching 2.28% ± 1.27% at day 3 and 0.38% ± 0.47% at day 7. Grafted cells were not detectable in the hindlimb of the host after 8 days of transplantation. There was a statistically significant difference in the reduction of the grafted-cell percentage in the limb after 3 days, compared to the first day ($n = 9$; $p < 0.05$). Meanwhile, a low survival cell rate was recorded at the lateral gastrocnemius site after 1 day (0.78% ± 0.83%; $n = 9$). This ratio increased significantly at day 4, reaching 2.29% ± 2.26%. At 7 days, this percentage dropped to nearly zero (0.51% ± 0.35%). After 7 days, there was no detectable presence of grafted cells in the LGS (Fig. 5B).

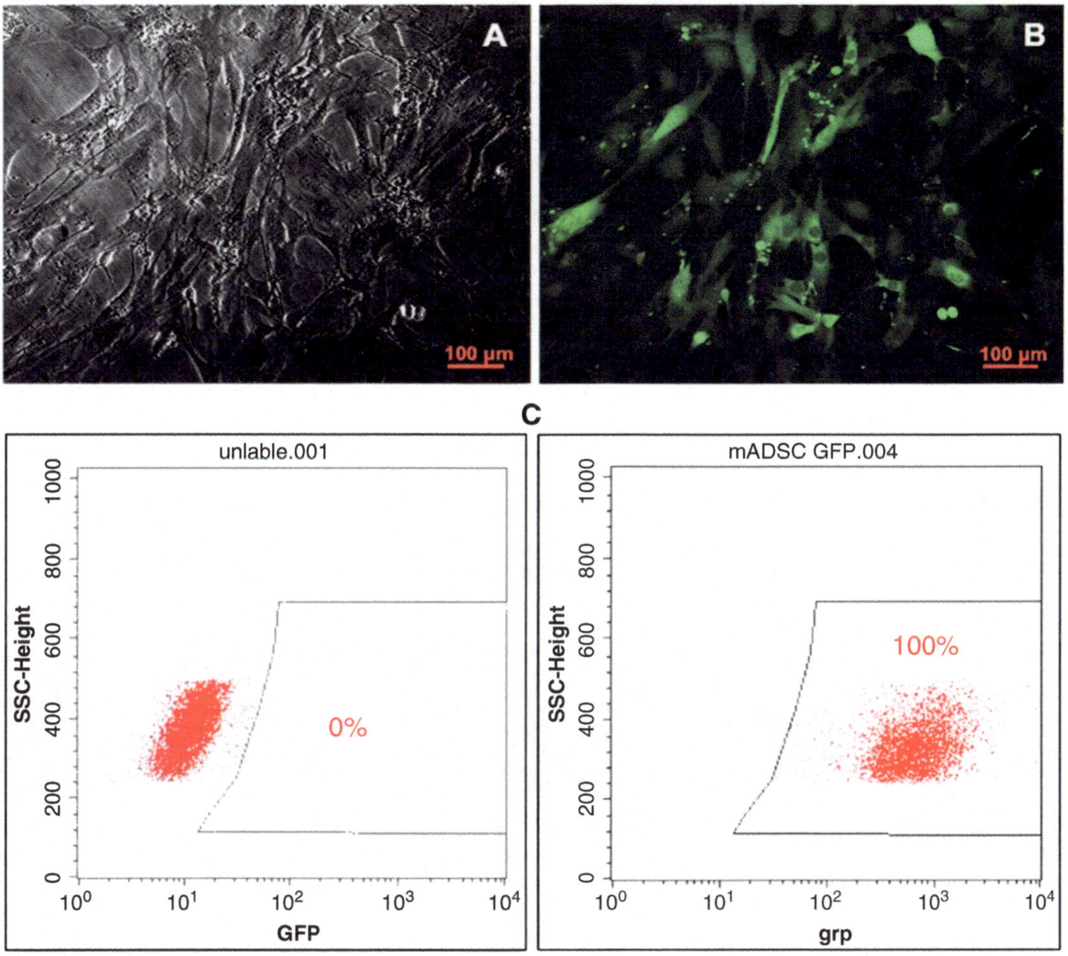

Fig. 2 *ADMSCs positive for green fluorescent protein (GFP) after transfected with GFP*. ADMSCs were captured in white field (**A**) and in fluorescent excitation for GFP (**B**). The expression of GFP in ADMSCs was confirmed by flow cytometry (**C**); 100% ADMSCs were positive

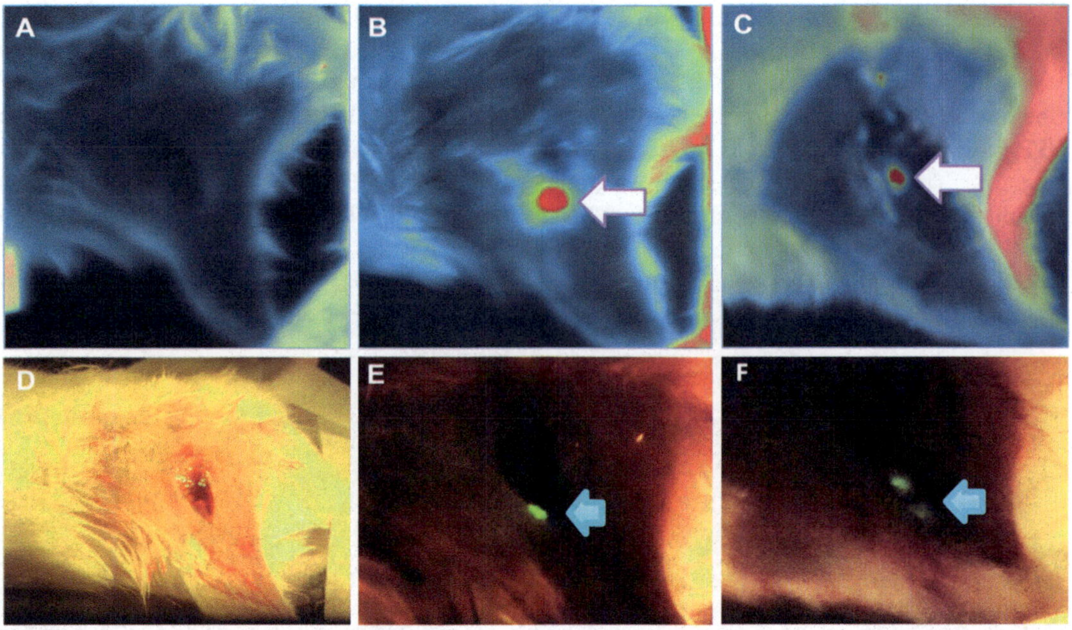

Fig. 3 *The detection of transplanted cells in the host.* Mouse hindlimb was scanned and imaged under green fluorescence at the PBS group (**A**) and at the ADMSC group immediately after (**B**) or 4 days after (**C**) cell transplantation. Observation with the naked eye through a fluorescence filter at the PBS group (**D**) and at the ADMSC group immediately after (**E**) or 1 h after (**F**) transplantation

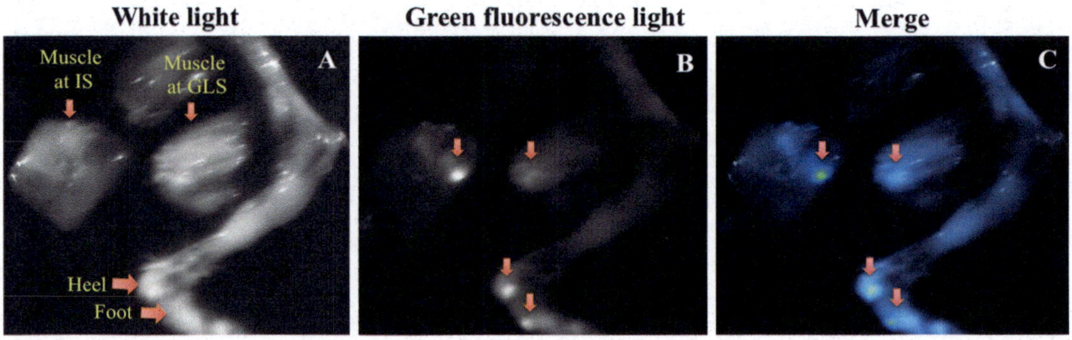

Fig. 4 *The movement of transplanted cells in the host.* Mouse hindlimb was visualized under white light (**A**) and green fluorescent light (**B**) and merged (**C**) after 4 days of ADMSC transplantation. Images show that the cells migrated to the heel and foot (*IS* injected site, *LGS* lateral gastrocnemius site)

3.4 Expression and Characteristics of Angiogenic Genes

The results showed that in our study, expression of several myogenic and angiogenic genes could be classified into four main groups (Table 2). The groups were as follows: Group I consists of angiogenic genes not expressed in the normal mice group compared to PBS injection or ADMSC injection group – *Flt-1*, *Flk-1*, *Ang-2*, *and VE-Cadherin*. Group II consists of angiogenic genes expressed in all groups – *HGF*, *CD31*, *TGF-beta*, *vWF*, *Ang-1*, *PGFRB*, and *ephrin-B4*. Group III consists of angiogenic

Table 2 Characteristics of angiogenic and myogenic genes

Group	Gene	Characteristic	Gene expression			
				Adult mouse		
					Ischemic hindlimb	
			Embryo	Normal	PBS injection	ADMSC injection
Internal control	*GAPDH*	Housekeeping gene; expressed in all cell types or tissues at high levels	+	+	+	+
Group I	*Flk-1*	Regulates endothelial cell migration and angiogenesis (Kappas et al. 2008)	+	–	+	+
	Flt-1	Chemotaxis; increases the vascular permeability, endothelial cells proliferation, and survival (Gille et al. 2001)	+	–	+	+
	ANG-2	Disrupts the connection between endothelial cells and extravascular cells (Fagiani and Christofori 2013)	+	–	+	+
	VE-cadherin	Controls and maintains endothelial cell contacts (Vestweber 2008)	+	–	+	+
Group II, Group II	*HGF*	Mitotic factors; morphology of tissue; regulates the migration of endothelial cells and plays an important role in angiogenesis (Ding et al. 2003)	+	+	+	+
	CD31	Important factor in the adhesion of endothelial cells during angiogenesis (Pusztaszeri et al. 2006)	+	+	+	+
	TGF-beta	Tissue morphology; induces apoptosis and angiogenesis (Ferrari et al. 2009a)	+	+	+	+
	vWF	Produced by endothelial cells; indicates the presence of endothelial cell (Pusztaszeri et al. 2006)	+	+	+	+
	Ang-1	Intermediates for the move, adhesion, and survival of endothelial cells (Fagiani and Christofori 2013)	+	+	+	+
	Ephrin-B4	Determines the formation of veins	+	+	+	+
Group III	*MMP-2*	Essential factor in the regulation of blood vessel branching activities (Newby 2005; van Hinsbergh et al. 2006)	+	–	–	–
	Ephrin-B2	Important factor in the formation of new blood vessels and vascular smooth muscle cell (Shin et al. 2001)	+	–	–	–
Group IV	*Myf5*	Regulates the homeostasis and regeneration of muscle cells; is expressed on myoblast (Gayraud-Morel et al. 2007)	+	+	+	+
	MyoD	Regulates the differentiation of myoblast (Gayraud-Morel et al. 2007)	+	–	+	+

(+) expression; (–) non-expression

genes expressed only in embryos, but not in normal mice or acute ischemic limb mice (with or without cell transplantation): *MMP-2* and *ephrin B-2*. Group IV includes the myogenic genes: *MyoD* and *Myf5*.

For Group I, although several angiogenic genes (such as *Flt-1*, *Flk-1*, and *Ang-2*) were not expressed in normal mice, the genes were expressed in the acute ischemic hindlimb mice, with or without transplantation of ADMSCs (Table 2). Strikingly, *Ang-2* expression in the ADMSC group at day 28 was significantly increased compared to day 3 after transplantation

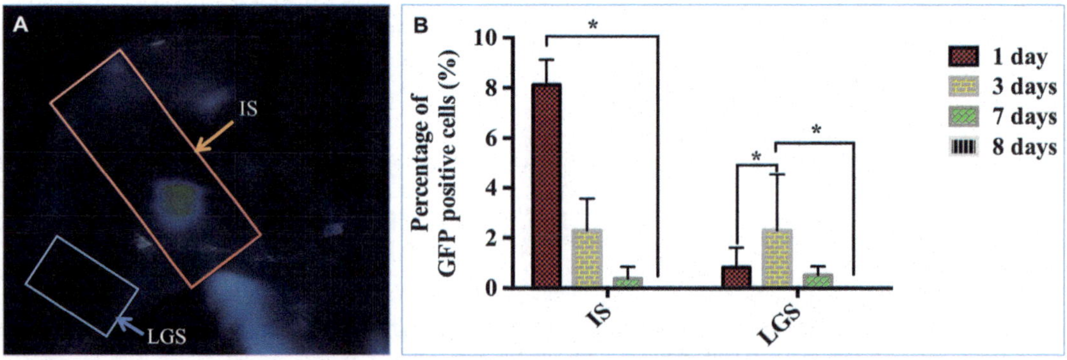

Fig. 5 *Percentage of detectable transplanted cells in the host*. Transplanted cells were collected from the hindlimb (**A**). Grafted cells were not detectable in the hindlimb of the host after 8 days of transplantation (**B**) (*$p < 0.05$, Sidak's multiple comparisons test; $n = 3$; repeated 3 times) (*IS* injected site, *LGS* lateral gastrocnemius site)

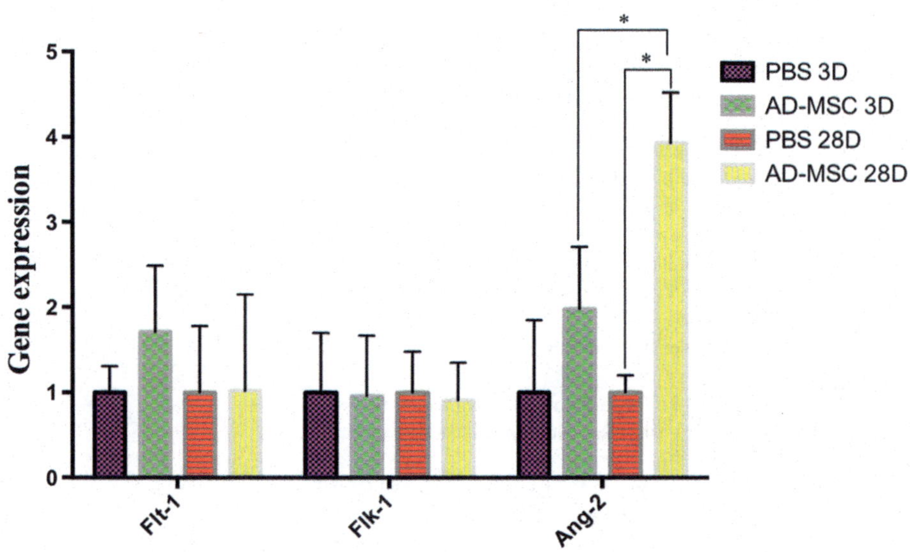

Fig. 6 *Gene expression of Flt-1, Flk-1, and Ang-2 after cell transplantation*. Expression of Ang-2 was significantly increased in the ADMSC group compared to the PBS group and increased in the ADMSC group at day 28 compared to day 3 (*$p < 0.05$, Tukey's multiple comparisons test; $n = 4$)

(p < 0.05) and compared to the PBS group ($p < 0.05$) (Fig. 6).

Meanwhile, *VE-cadherin* expression was not detected in normal limb but was found in acute ischemic limb in the PBS group at day 3. However, *VE-cadherin* expression was not detectable anymore at day 28. Conversely, *VE-cadherin* expression in the ADMSC group was detectable on both days 3 and 28.

For Group II (Table 3, Fig. 7), the expression of *HGF*, *CD31*, *and TGF-β* were significantly increased after 3 days of cell transplantation compared to normal mice ($p < 0.05$) and significantly decreased after 28 days ($p < 0.05$) except

Table 3 Gene expression analysis showing an increase of *HGF, CD31, TGF-ß, vWF, Ang-1, and ephrin-B4* in ADMSC-transplanted hindlimb ischemic mice compared to normal mice

	Gene expression (ADMSC group compared to normal group)			
Genes	PBS-3D	ADMSC-3D	PBS-28D	ADMSC-28D
HGF	1.09 ± 0.25	42.62 ± 12.63[a]	0.12 ± 0.09	13.77 ± 3.28[b]
CD31	3.94 ± 3.59	9.44 ± 10.81[a]	0.41 ± 0.29	0.30 ± 0.28[b]
TGF-B	3.84 ± 1.26	9.63 ± 5.33[a]	0.43 ± 0.05	1.05 ± 1.11[b]
vWF	0.46 ± 0.33	3.87 ± 1.94	0.19 ± 0.26	0.86 ± 0.48
Ang-1	0.66 ± 0.61	0.25 ± 0.09	0.23 ± 0.08	0.77 ± 0.22
Eph-B4	2.15 ± 2.05	1.22 ± 0.62	0.34 ± 0.08	1.33 ± 0.80

[a]Statistically significant difference compared to the normal group
[b]Statistically significant difference compared to the ADMSC group at day 3
Tukey's multiple comparisons test, $p < 0.05$; $n = 4$

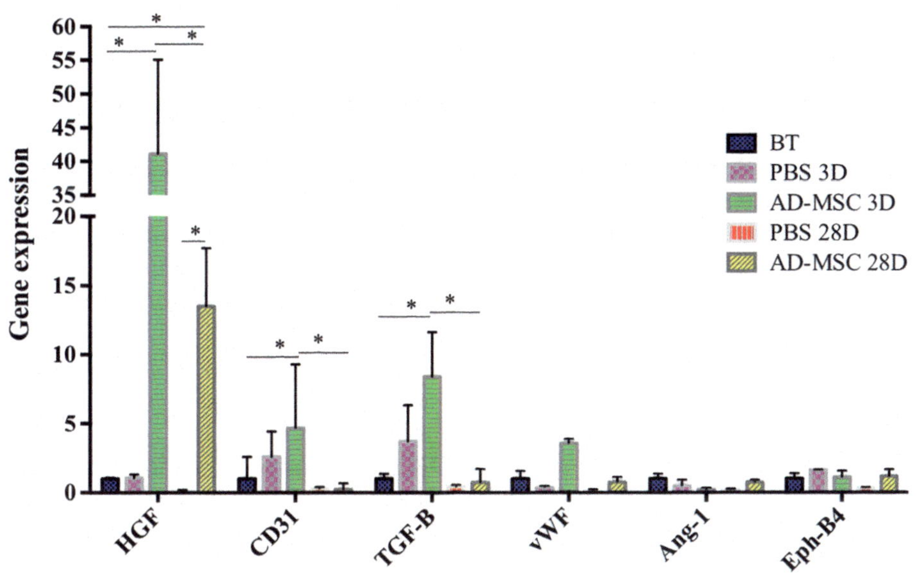

Fig. 7 *Gene expression of HGF, CD31, TGF-beta, vWF, Ang-1, and ephrin-B4 after cell transplantation.* *$p < 0.05$, Tukey's multiple comparisons test; $n = 4$

for *HGF*. Expression of *HGF* was significantly higher in the ADMSC group at day 28 compared to the normal group. Meanwhile, there were no significant changes in expression of these genes in the non-treated group compared to the normal group ($p > 0.05$). Moreover, *vWF, Ang-1*, and *ephrin-b4* expression showed no significant changes in both the cell-transplanted and non-treated mice.

For Group III, *MMP-2* and *ephrin-B2* expression was only observed in embryos but not in normal or acute hindlimb ischemic adult mice (with or without cell transplantation).

For Group IV, *MyoD* was expressed only in the PBS and ADMSC groups but not in the normal group. On the other hand, gene expression increased significantly in the ADMSC group at day 28 (4.4 ± 2.0 fold; $n = 4$) compared to day 3 (1.7 ± 1.2 fold; $n = 4$) ($p < 0.05$) and compared to the PBS group ($p < 0.05$) (Fig. 8).

Meanwhile, *Myf5* was expressed in all groups. At day 3, *Myf5* expression was higher in both the PBS and ADMSC groups compared to the normal group (3.7 ± 1.5 and 7.7 ± 6.9 fold, respectively). At day 28, the expression increased to 3.4 ± 2.2 fold and 4.4 ± 2.3 fold for those

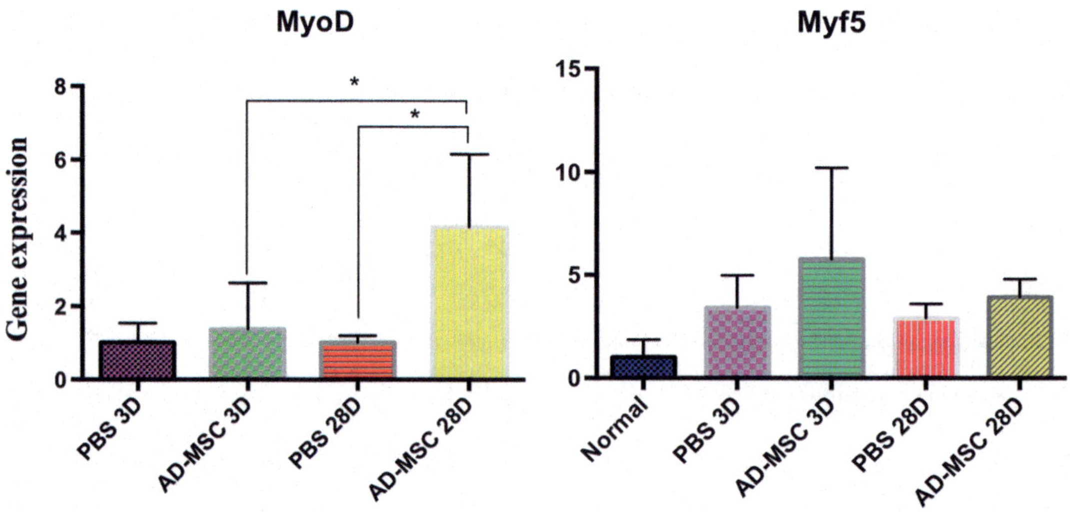

Fig. 8 *Expression of MyoD and Myf5 after cell transplantation* (*$p < 0.05$, Tukey's multiple comparisons test; $n = 4$)

groups, respectively. However, the differences between the groups and among the time points did not differ significantly ($p > 0.05$) (Fig. 8).

4 Discussion

The migration and/or molecular mechanisms of cells during healing of damaged vascular or muscle tissue are emerging areas of research interest worldwide. Due to signals during injury and by the control of autocrine and/or paracrine factors, intrinsic stem cells or transplanted stem cells receive specific signals to initiate the repair process. They migrate to the damaged sites to participate in wound healing. The complex migration process of stem cells is driven by a large number of cytokines, adhesion molecules, and growth factors, all of which are necessary for controlling functional tissue regeneration safely and effectively (Kalka et al. 2000).

In this study, GFP-transduced ADMSCs after transplantation were detectable along the acute ischemic hindlimb, as evidenced by the separation of the light spots and by the reduction of their size at the cell-transplanted position. However, the reduction of the light spot size (i.e., the decrease in concentration of transplanted cells) may be due to necrosis of cells in the host.

Two mechanisms may cause necrosis of cells in vitro: cell detachment from culture surfaces and removal of growth factors. Necrosis due to lack of adhesion factors is known as anoikis (Frisch and Francis 1994) and involves the activation of caspase-3 (Yuan and Yankner 2000). Single cells prepared before transplantation can lose their ECM-integrin interactions, and as a result, apoptosis is triggered (Yuan and Yankner 2000). Cell survival was also limited by the environment of the damaged region. Injury at the graft area may be due to the needle, the removal of nutrients, the shock caused by oxidation, and/or a combination of these factors (Bakshi et al. 2005).

On the other hand, the reduction of cell density could be due to graft rejection by the host. In this study, allogeneic MSCs were used to treat disease. The ability to reject allogeneic graft in the host has been reported in many previous studies (Ball and Egeler 2008). However, mesenchymal stem cells possess potent immunosuppressive factors as reported in several studies (Lee et al. 2002; Ning et al. 2008; Vianello and Dazzi 2008). In this study, the untreated immunosuppressed mice before transplantation did not show any active state of immune response. The proportion of live mice after cell transplantation was 100% over a 6-month period.

This indicates that a reduction in cell density due to destruction by the host is not the main cause.

Another possible reason for the reduction of light-emitting cells in the grafts might be due to the movement of cells to other locations, which can lead to dispersed signals. At a certain degree of dispersion, the identification of fluorescence by the iBox system becomes inferior, and the existence of cells with sparse gathering can cause the system not to recognize or detect the cells. During the injury period, host cells secrete various factors which have a positive impact on the resident MSCs. The migratory and resident MSCs are dependent partly on chemokine receptors. Like white blood cells, MSCs express multiple receptors and cell adhesion molecules, including selectins and integrins. These factors induce migrant and resident transplanted cells to move to the target tissue. At the damaged tissue location, inflammatory reactions occur. For instance, TNF-alpha and IL-1 (Ruster et al. 2006; Shi et al. 2007) are released and activate endothelial cells to express VCAM-1 and activate MSCs. As a result, MSCs express integrin β1 and/or integrin α4/β1 and CXCR4 (Krampera et al. 2006; Ruster et al. 2006; Shi et al. 2007). At this time, MSCs migrate to endothelial cells then pass through the endothelium via interaction with the extracellular matrix and cell-cell interactions. By this process, MSCs can participate in repairing the tissue damage. When blood flow is interrupted, the blood supply is reduced, thereby inducing tissue injury. Signals from damaged tissues in response to receptors on MSCs can trigger the migration of cells (Sohni and Verfaillie 2013). In our study, the survival of cells in the host was decreased over time at the cell-transplanted area but increased at non-transplanted areas (e.g., foot). The results clearly show that grafted cells were capable of migrating to the area under the cut blood vessels.

Flt-1, also known as VEGF receptor 1 (VEGFR-1), was previously known as a limited angiogenic factor (Ferrara et al. 2003; Shibuya 2006). However, the role of flt-1 in angiogenesis has been controversial. Several reports have shown a positive impact of flt-1 on the migration of endothelial cells and the formation of new blood vessels (Gerber et al. 1997; Olofsson et al. 1998). Kappas et al. have confirmed that *flt-1* can adjust FLK-1/KDR signal, regulate the migration of endothelial cells, and impact angiogenesis (Kappas et al. 2008). Other studies using flt-1 knockout mice have shown that flt-1 stimulates the proliferation of endothelial cells. Moreover, other studies have suggested that flt-1 is essential for forming new blood vessels (Fong et al. 1999; Park et al. 1994). Mishi et al. have also reported that flt-1 regulates Akt, and together they play an important role in the maintenance of endothelial cell integrity and stability of vascular structures (Nishi et al. 2008). Flt-1 is associated not only with chemotaxis and increase of vascular permeability but also with the proliferation and survival of endothelial cells (Gille et al. 2001).

Fagiani et al. have shown that *Ang-2*, on the other hand, interrupts the interactions between endothelial cells and pericytes and induces apoptosis of pericytes. This leads to loose or degenerative vascular structures and triggers angiogenesis. When combined with *VEGF*, *Ang-2* induces the formation of new blood vessels (Fagiani and Christofori 2013). In this study, *Flt-1* and *Flk-1* were not expressed in normal mice but were activated in hindlimb ischemic mice, with or without stem cell treatment. This suggests that *VEGF* is expressed. The presence of *VEGF* in combination with the *Ang-2* expression showed that there is induction of new blood vessel formation.

In addition, *VE-cadherin* is a specific adhesion molecule for endothelial cells and is located at intercellular junctions between them. *VE-cadherin* is an important factor for controlling and maintaining the interactions between endothelial cells. It also regulates many different processes, such as cell proliferation, apoptosis, and mediation of VEGF receptors (Vestweber 2008).

CD31 (or *PECAM-1*) is a transmembrane glycoprotein present on white blood cells, macrophages, platelets, endothelial cells, and stem cells. Recently, CD31 has been recognized as an essential factor in angiogenesis (DeLisser et al. 1997; Matsumura et al. 1997; Zhou et al.

1999). In a study by Hellingman et al., CD31 expression was significantly higher in acute hindlimb ischemic mouse than in normal mouse. Meanwhile, the capillary density increased in proportion to *CD31* expression on cells (Hellingman et al. 2010). In our study, the expression of *CD31* increased, similar to the studies by Hellingman et al.

Other angiogenic genes (*HGF, TGF-β, vWF, Ang-1*, and *ephrin-B4*) were assessed in our study. *HGF* is identified as a member of the vascular growth factor family (Morishita et al. 2004). TGF-β is regulated by tissue morphology; half of TGF-β-deficient mice died in utero in mothers due to lack of new blood vessel formation (Ferrari et al. 2009). In this study, the expression of HGF and TGF-β was significantly increased in the ADMSC group compared to the PBS and normal groups. This shows that stem cells and endothelial cells might be more strongly activated in those groups. This also implies that the damage repair process was stronger in the cell-transplanted group. In our study, *vWF, Ang-1*, and *ephrin-B4* did not differ significantly among the groups. However, *Flk-1, Flt-1, Ang-2, CD31, HGF, and TGF-β* are all essential for the process of angiogenesis.

Myf5 and *MyoD* are genes identified to be myogenic and promote the fate of skeletal muscle cells after birth. As well, they are associated with repair of tissue damage. In adult, *MyoD* regulates myoblast differentiation, while *Myf5* regulates homeostasis and regeneration of muscle cells (Gayraud-Morel et al. 2007). Our study showed that activation of these genes in muscle regeneration is triggered after acute ischemia. Moreover, after cell transplantation, *MyoD* was strongly expressed at day 28, which indicates that ADMSC transplantation had a positive impact on repairing ischemic tissue.

5 Conclusion

The results from our preliminary study show that ADMSCs participate in the healing and treatment of acute ischemic hindlimb. Several mechanisms are used by ADMSCs, including movement of transplanted cells to affected positions (foot, heel, and lower thigh) where blockage of blood vessels often occurs. The ADMSCs can persist at these locations for about 7 days. In our study, the ADMSCs can repair tissue damage by increasing the expression of several angiogenesis-related genes, including *VE-cadherin, HGF, CD31, TGF-β, Ang-2, Flt-1, Flk-1*, and *MyoD*.

Acknowledgments This research was funded by the University of Science, Vietnam National University, Ho Chi Minh City, Vietnam, under grant number T2016-20.

Competing Interests The authors declare that no competing interests exist.

Authors' Contributions NBV and NPK were responsible for suggesting the idea for this study, creating the experiment design, analyzing the data, writing the result, discussing, and preparing the figures. HLTN was responsible for performing the RT-PCR analysis and writing the introduction and methods. LTP was responsible for performing the ADMSCs cultures. TTTD was responsible for creating GFP-ADMSC and preparing muscle cell for flow cytometry analysis. NBV and VTT were responsible for analyzing the flow cytometry and revising the manuscript. All authors read and approved the manuscript.

References

Bakshi, A., Keck, C. A., Koshkin, V. S., LeBold, D. G., Siman, R., Snyder, E. Y., & McIntosh, T. K. (2005). Caspase-mediated cell death predominates following engraftment of neural progenitor cells into traumatically injured rat brain. *Brain Research, 1065*, 8–19.

Ball, L. M., & Egeler, R. M. (2008). Acute GvHD: Pathogenesis and classification. *Bone Marrow Transplantation, 41*, S58–S64.

Baraniak, P. R., & McDevitt, T. C. (2010). Stem cell paracrine actions and tissue regeneration. *Regenerative Medicine, 5*, 121–143.

Buschmann, I., & Schaper, W. (1999). Arteriogenesis versus angiogenesis: Two mechanisms of vessel growth. *Physiology, 14*, 121–125.

Conway, E. M., Collen, D., & Carmeliet, P. (2001). Molecular mechanisms of blood vessel growth. *Cardiovascular Research, 49*, 507–521.

Cui, Z., Zhou, H., He, C., Wang, W., Yang, Y., & Tan, Q. (2015). Upregulation of Bcl-2 enhances secretion of growth factors by adipose-derived stem cells deprived of oxygen and glucose. *Bioscience Trends, 9*, 122–128.

DeLisser, H. M., Christofidou-Solomidou, M., Strieter, R. M., Burdick, M. D., Robinson, C. S., Wexler, R. S., Kerr, J. S., Garlanda, C., Merwin, J. R., & Madri, J. A. (1997). Involvement of endothelial PECAM-1/CD31 in angiogenesis. *The American Journal of Pathology, 151*, 671.

Ding, S., Merkulova-Rainon, T., Han, Z. C., & Tobelem, G. (2003). HGF receptor up-regulation contributes to the angiogenic phenotype of human endothelial cells and promotes angiogenesis in vitro. *Blood, 101*, 4816–4822.

Dvorak, H. F., Brown, L. F., Detmar, M., & Dvorak, A. M. (1995). Vascular permeability factor/vascular endothelial growth factor, microvascular hyperpermeability, and angiogenesis. *The American Journal of Pathology, 146*, 1029–1039.

Fagiani, E., & Christofori, G. (2013). Angiopoietins in angiogenesis. *Cancer Letters, 328*, 18–26.

Ferrara, N., Gerber, H.-P., & LeCouter, J. (2003). The biology of VEGF and its receptors. *Nature Medicine, 9*, 669–676.

Ferrari, G., Cook, B. D., Terushkin, V., Pintucci, G., & Mignatti, P. (2009). Transforming growth factor-beta 1 (TGF-β1) induces angiogenesis through vascular endothelial growth factor (vegf)-mediated apoptosis. *Journal of Cellular Physiology, 219*, 449–458.

Fong, G.-H., Zhang, L., Bryce, D.-M., & Peng, J. (1999). Increased hemangioblast commitment, not vascular disorganization, is the primary defect in flt-1 knock-out mice. *Development, 126*, 3015–3025.

Fouraschen, S. M. G., Wolf, J. H., van der Laan, L. J. W., de Ruiter, P. E., Hancock, W. W., van Kooten, J. P., Verstegen, M. M. A., Olthoff, K. M., & de Jonge, J. (2015). Mesenchymal stromal cell-derived factors promote tissue repair in a small-for-size ischemic liver model but do not protect against early effects of ischemia and reperfusion injury. *Journal of Immunology Research, 2015*, 202975. doi:10.1155/2015/202975. Epub 2015 Aug 24.

Frisch, S. M., & Francis, H. (1994). Disruption of epithelial cell-matrix interactions induces apoptosis. *The Journal of Cell Biology, 124*, 619–626.

Gayraud-Morel, B., Chrétien, F., Flamant, P., Gomès, D., Zammit, P. S., & Tajbakhsh, S. (2007). A role for the myogenic determination gene Myf5 in adult regenerative myogenesis. *Developmental Biology, 312*, 13–28.

Gerber, H.-P., Condorelli, F., Park, J., & Ferrara, N. (1997). Differential transcriptional regulation of the two vascular endothelial growth factor receptor genes Flt-1, but not Flk-1/KDR, is up-regulated by hypoxia. *Journal of Biological Chemistry, 272*, 23659–23667.

Gerhardt, H., & Betsholtz, C. (2003). Endothelial-pericyte interactions in angiogenesis. *Cell and Tissue Research, 314*, 15–23.

Gille, H., Kowalski, J., Li, B., LeCouter, J., Moffat, B., Zioncheck, T. F., Pelletier, N., & Ferrara, N. (2001). Analysis of biological effects and signaling properties of Flt-1 (VEGFR-1) and KDR (VEGFR-2). A reassessment using novel receptor-specific vascular endothelial growth factor mutants. *The Journal of Biological Chemistry, 276*, 3222–3230.

Hayashi, S.-I., Morishita, R., Nakamura, S., Yamamoto, K., Moriguchi, A., Nagano, T., Taiji, M., Noguchi, H., Matsumoto, K., Nakamura, T., et al. (1999). Potential role of hepatocyte growth factor, a novel angiogenic growth factor, in peripheral arterial disease. *Downregulation of HGF in Response to Hypoxia in Vascular Cells, 100*, II-301–II-308.

Hellingman, A. A., Bastiaansen, A. J. N. M., de Vries, M. R., Seghers, L., Lijkwan, M. A., Löwik, C. W., Hamming, J. F., & Quax, P. H. A. (2010). Variations in surgical procedures for hind limb Ischaemia mouse models result in differences in collateral formation. *European Journal of Vascular and Endovascular Surgery, 40*, 796–803.

Hertig, A. T. (1935). Angiogenesis in the early human chorion and in the primary placenta of the macaque monkey. *Contrib Embryol, 25*, 39–81.

Iwase, T., Nagaya, N., Fujii, T., Itoh, T., Murakami, S., Matsumoto, T., Kangawa, K., & Kitamura, S. (2005). Comparison of angiogenic potency between mesenchymal stem cells and mononuclear cells in a rat model of hindlimb ischemia. *Cardiovascular Research, 66*, 543–551.

Kalka, C., Masuda, H., Takahashi, T., Kalka-Moll, W. M., Silver, M., Kearney, M., Li, T., Isner, J. M., & Asahara, T. (2000). Transplantation of ex vivo expanded endothelial progenitor cells for therapeutic neovascularization. *Proceedings of the National Academy of Sciences of the United States of America, 97*, 3422–3427.

Kamihata, H., Matsubara, H., Nishiue, T., Fujiyama, S., Tsutsumi, Y., Ozono, R., Masaki, H., Mori, Y., Iba, O., Tateishi, E., et al. (2001). Implantation of bone marrow mononuclear cells into ischemic myocardium enhances collateral perfusion and regional function via side supply of angioblasts, angiogenic ligands, and cytokines. *Circulation, 104*, 1046–1052.

Kappas, N. C., Zeng, G., Chappell, J. C., Kearney, J. B., Hazarika, S., Kallianos, K. G., Patterson, C., Annex, B. H., & Bautch, V. L. (2008). The VEGF receptor Flt-1 spatially modulates Flk-1 signaling and blood vessel branching. *The Journal of Cell Biology, 181*, 847–858.

Karamysheva, A. F. (2008). Mechanisms of angiogenesis. *Biochemistry Biokhimiia, 73*, 751–762.

Kim, N., & Cho, S. G. (2015). New strategies for overcoming limitations of mesenchymal stem cell-based immune modulation. *International Journal of Stem Cells, 8*, 54–68.

Krampera, M., Pizzolo, G., Aprili, G., & Franchini, M. (2006). Mesenchymal stem cells for bone, cartilage, tendon and skeletal muscle repair. *Bone, 39*, 678–683.

Lee, S. T., Jang, J. H., Cheong, J. W., Kim, J. S., Maemg, H. Y., Hahn, J. S., Ko, Y. W., & Min, Y. H. (2002). Treatment of high-risk acute myelogenous leukaemia by myeloablative chemoradiotherapy followed by co-infusion of T cell-depleted haematopoietic stem cells and culture-expanded marrow mesenchymal stem cells from a related donor with one fully mismatched human leucocyte antigen haplotype. *British Journal of Haematology, 118*, 1128–1131.

Li, D., Zhang, M., Zhang, Q., Wang, Y., Song, X., & Zhang, Q. (2015). Functional recovery after acute intravenous administration of human umbilical cord mesenchymal stem cells in rats with cerebral ischemia-reperfusion injury. *Intractable & Rare Diseases Research, 4*, 98–104.

Marti, H. J. H., Bernaudin, M., Bellail, A., Schoch, H., Euler, M., Petit, E., & Risau, W. (2000). Hypoxia-induced vascular endothelial growth factor expression precedes Neovascularization after cerebral ischemia. *The American Journal of Pathology, 156*, 965–976.

Matsumura, T., Wolff, K., & Petzelbauer, P. (1997). Endothelial cell tube formation depends on cadherin 5 and CD31 interactions with filamentous actin. *Journal of Immunology (Baltimore, Md: 1950), 158*, 3408–3416

Morishita, R., Aoki, M., Hashiya, N., Yamasaki, K., Kurinami, H., Shimizu, S., Makino, H., Takesya, Y., Azuma, J., & Ogihara, T. (2004). Therapeutic angiogenesis using hepatocyte growth factor (HGF). *Current Gene Therapy, 4*, 199–206.

Nakamura, T., Mizuno, S., Matsumoto, K., Sawa, Y., Matsuda, H., & Nakamura, T. (2000). Myocardial protection from ischemia/reperfusion injury by endogenous and exogenous HGF. *The Journal of Clinical Investigation, 106*, 1511–1519.

Newby, A. C. (2005). Dual role of matrix metalloproteinases (Matrixins) in intimal thickening and atherosclerotic plaque rupture. *Physiological Reviews, 85*, 1–31.

Ning, H., Yang, F., Jiang, M., Hu, L., Feng, K., Zhang, J., Yu, Z., Li, B., Xu, C., Li, Y., et al. (2008). The correlation between cotransplantation of mesenchymal stem cells and higher recurrence rate in hematologic malignancy patients: Outcome of a pilot clinical study. *Leukemia, 22*, 593–599.

Nishi, J.-I., Minamino, T., Miyauchi, H., Nojima, A., Tateno, K., Okada, S., Orimo, M., Moriya, J., Fong, G.-H., & Sunagawa, K. (2008). Vascular endothelial growth factor receptor-1 regulates postnatal angiogenesis through inhibition of the excessive activation of Akt. *Circulation Research, 103*, 261–268.

Nishida, N., Yano, H., Nishida, T., Kamura, T., & Kojiro, M. (2006). Angiogenesis in cancer. *Vascular Health and Risk Management, 2*, 213–219.

Olofsson, B., Korpelainen, E., Pepper, M. S., Mandriota, S. J., Aase, K., Kumar, V., Gunji, Y., Jeltsch, M. M., Shibuya, M., & Alitalo, K. (1998). Vascular endothelial growth factor B (VEGF-B) binds to VEGF receptor-1 and regulates plasminogen activator activity in endothelial cells. *Proceedings of the National Academy of Sciences, 95*, 11709–11714.

Park, J. E., Chen, H. H., Winer, J., Houck, K. A., & Ferrara, N. (1994). Placenta growth factor. Potentiation of vascular endothelial growth factor bioactivity, in vitro and in vivo, and high affinity binding to Flt-1 but not to Flk-1/KDR. *Journal of Biological Chemistry, 269*, 25646–25654.

Pipino, C., & Pandolfi, A. (2015). Osteogenic differentiation of amniotic fluid mesenchymal stromal cells and their bone regeneration potential. *World Journal of Stem Cells, 7*, 681–690.

Pusztaszeri, M. P., Seelentag, W., & Bosman, F. T. (2006). Immunohistochemical expression of endothelial markers CD31, CD34, von Willebrand factor, and Fli-1 in normal human tissues. *The Journal of Histochemistry and Cytochemistry: Official Journal of the Histochemistry Society, 54*, 385–395.

Rennert, R. C., Sorkin, M., Garg, R. K., & Gurtner, G. C. (2012). Stem cell recruitment after injury: Lessons for regenerative medicine. *Regenerative Medicine, 7*, 833–850.

Ruster, B., Gottig, S., Ludwig, R. J., Bistrian, R., Muller, S., Seifried, E., Gille, J., & Henschler, R. (2006). Mesenchymal stem cells display coordinated rolling and adhesion behavior on endothelial cells. *Blood, 108*, 3938–3944.

Shi, M., Li, J., Liao, L., Chen, B., Li, B., Chen, L., Jia, H., & Zhao, R. C. (2007). Regulation of CXCR4 expression in human mesenchymal stem cells by cytokine treatment: Role in homing efficiency in NOD/SCID mice. *Haematologica, 92*, 897–904.

Shibuya, M. (2006). Vascular endothelial growth factor receptor-1 (VEGFR-1/Flt-1): A dual regulator for angiogenesis. *Angiogenesis, 9*, 225–230. discussion 231.

Shin, D., Garcia-Cardena, G., Hayashi, S.-I., Gerety, S., Asahara, T., Stavrakis, G., Isner, J., Folkman, J., Gimbrone, M. A., & Anderson, D. J. (2001). Expression of EphrinB2 identifies a stable genetic difference between arterial and venous vascular smooth muscle as well as endothelial cells, and marks subsets of microvessels at sites of adult neovascularization. *Developmental Biology, 230*, 139–150.

Smart, N., & Riley, P. R. (2008). The stem cell movement. *Circulation Research, 102*, 1155–1168.

Sohni, A., & Verfaillie, C. M. (2013). Mesenchymal stem cells migration homing and tracking. *Stem Cells International, 2013*, 8.

Strem, B. M., Hicok, K. C., Zhu, M., Wulur, I., Alfonso, Z., Schreiber, R. E., Fraser, J. K., & Hedrick, M. H. (2005). Multipotential differentiation of adipose tissue-derived stem cells. *The Keio Journal of Medicine, 54*, 132–141.

Sun, C. K., Leu, S., Hsu, S. Y., Zhen, Y. Y., Chang, L. T., Tsai, C. Y., Chen, Y. L., Chen, Y. T., Tsai, T. H., Lee, F. Y., et al. (2015). Mixed serum-deprived and normal adipose-derived mesenchymal stem cells against acute lung ischemia-reperfusion injury in rats. *American Journal of Translational Research, 7*, 209–231.

Toma, C., Pittenger, M. F., Cahill, K. S., Byrne, B. J., & Kessler, P. D. (2002). Human mesenchymal stem cells differentiate to a cardiomyocyte phenotype in the adult murine heart. *Circulation, 105*, 93–98.

Ucuzian, A. A., Gassman, A. A., East, A. T., & Greisler, H. P. (2010). Molecular mediators of angiogenesis. *Journal of Burn Care & Research: Official Publication of the American Burn Association, 31*, 158.

van Hinsbergh, V. W., Engelse, M. A., & Quax, P. H. (2006). Pericellular proteases in angiogenesis and vasculogenesis. *Arteriosclerosis, Thrombosis, and Vascular Biology, 26*, 716–728.

Van Pham, P., Vu, N. B., Phan, N. L.-C., Le, D. M., Truong, N. C., Truong, N. H., Bui, K. H.-T., & Phan, N. K. (2014). Good manufacturing practice-compliant isolation and culture of human adipose-derived stem cells. *Biomedical Research and Therapy, 1*, 133–141.

Vestweber, D. (2008). VE-cadherin: The major endothelial adhesion molecule controlling cellular junctions and blood vessel formation. *Arteriosclerosis, Thrombosis, and Vascular Biology, 28*, 223–232.

Vianello, F., & Dazzi, F. (2008). Mesenchymal stem cells for graft-versus-host disease: A double edged sword? *Leukemia, 22*, 463–465.

Vu, N. B., Trinh, V. N.-L., Phi, L. T., Bui, A. N.-T., Ta, V. T., Phan, N. K., & Van Pham, P. (2013). Optimizing the procedure for preparing immune- deficient hindlimb ischemia mouse model. *Vietnam Joural of Physiology, 17*, 27–36.

Yuan, J., & Yankner, B. A. (2000). Apoptosis in the nervous system. *Nature, 407*, 802–809.

Zentilin, L., Tafuro, S., Zacchigna, S., Arsic, N., Pattarini, L., Sinigaglia, M., & Giacca, M. (2006). Bone marrow mononuclear cells are recruited to the sites of VEGF-induced neovascularization but are not incorporated into the newly formed vessels. *Blood, 107*, 3546–3554.

Zhou, Z., Christofidou-Solomidou, M., Garlanda, C., & DeLisser, H. M. (1999). Antibody against murine PECAM-1 inhibits tumor angiogenesis in mice. *Angiogenesis, 3*, 181–188.

Zhuang, H., Zhang, X., Zhu, C., Tang, X., Yu, F., Shang, G. W., & Cai, X. (2016). Molecular mechanisms of PPAR-gamma governing MSC osteogenic and adipogenic differentiation. *Current Stem Cell Research & Therapy, 11*, 255–264.

Zygmunt, M., Herr, F., Munstedt, K., Lang, U., & Liang, O. D. (2003). Angiogenesis and vasculogenesis in pregnancy. *European Journal of Obstetrics, Gynecology, and Reproductive Biology, 110*(Suppl 1), S10–S18.

Advs Exp. Medicine, Biology - Innovations in Cancer Research and Regenerative Medicine (2018) 1: 19–28
DOI 10.1007/5584_2017_67

Published online: 8 July 2017

Stem Cell Therapy for Fanconi Anemia

Qing-Shuo Zhang

Abstract

Stem cell therapy is the administration of stem cells to a patient to treat or prevent a disease. Since stem cells possess the long-term self-renewal capacity and provide daughter cells that differentiate into the specialized cells of each tissue, stem cell therapy will theoretically improve the disease condition for the lifetime of the patient. As the most widely used stem cell therapy, bone marrow transplantation is the treatment of choice for many kinds of blood disorders, including anemias, leukemias, lymphomas, and rare immunodeficiency diseases. For the fatal genetic blood disorder Fanconi anemia, allogeneic bone marrow transplantation has remained the only curative treatment. But the recent advances in stem cell and gene therapy fields may provide promising opportunities for an alternative or even better management of Fanconi anemia. Many of these new ideas and opportunities are also useful for treating other blood diseases that affect hematopoietic stem cells, such as sickle cell anemia, severe combined immunodeficiencies, and beta-thalassemias. In this chapter, these advances along with their challenges and limitations will be thoroughly discussed.

Keywords

Fanconi anemia • Hematopoietic stem cell • Stem cell therapy

Q. S. Zhang (✉)
Oregon Stem Cell Center, Department of Pediatrics, Oregon Health & Science University, 3181 SW Sam Jackson Park Road, Portland, OR 97239, USA
e-mail: zhangqi@ohsu.edu

Abbreviations

AAV	Adeno-associated virus
FA	Fanconi anemia
GVHD	Graft-versus-host disease
HLA	Human leukocyte antigens
HSCT	Hematopoietic stem cell transplantation
HSPC	Hematopoietic stem and progenitor cell
iPS	Induced pluripotent stem cell

1 Introduction

Fanconi anemia (FA) is a rare genetic disorder characterized by birth defects, bone marrow failure, and cancer susceptibility (Auerbach 2009). Recent research studies have revealed that FA is caused by an inherited defect in the ability to repair interstrand DNA crosslinks in cells. Bi-allelic mutations in any one of 21 causative genes (FANCA, FANCB, FANCC, FANCD1/BRCA2, FANCD2, FANCE, FANCF, FANCG/XRCC9, FANCI, FANCL/PHF9/Pog, FANCJ/BRIP1/BACH1, FANCM/Hef, FANCN/PALB2, FANCO/RAD51C, FANCP/SLX4, FANCQ/XPF/ERCC4, FANCR/RAD51, FANCS/BRCA1, FANCT/UBE2T, FANCU/XRCC2, and FANCV/MAD2L2/REV7) can cause this disease (Dong et al. 2015). All of these 21 FA genes are believed to function in a common cellular signaling pathway that responds to DNA damage and maintains genome stability (Kim and D'Andrea 2012).

The primary cause of morbidity and mortality in FA patients is progressive bone marrow failure, which shortens the average life span of the patients (Kutler et al. 2003). FA patients have fewer hematopoietic stem/progenitor cells (HSPCs) in their bone marrow and suffer from additional progressive loss of HSPCs due to an exacerbated p53/p21 DNA damage response (Ceccaldi et al. 2011, 2012; Kelly et al. 2007). Most patients develop marrow dysfunction within the first decade of life. The symptoms range from mild cytopenia in any lineage, with thrombocytopenia as the most common type, to severe aplastic anemia (Shimamura and Alter 2010).

Pharmacological therapy to manage hematologic abnormalities for FA patients is limited and has remained largely unchanged for over 30 years. Androgen therapy normalizes blood counts in ~50–70% of FA patients and also works for other forms of aplastic anemia (Dokal 2003). The most commonly used androgen is oxymetholone, which is an anabolic-androgenic steroid and a synthetic derivative of testosterone (Shimamura and Alter 2010). The beneficial effects of androgens are most pronounced in platelets and red blood cells. Androgens also improve neutrophil counts in some patients. However, androgen therapy is not a permanent rescue of hematopoiesis. Many FA patients eventually become refractory to androgen therapy even if their initial response was good. It has been proposed that hematopoietic stem cells in these patients may have been completely depleted at later stages of androgen therapy (Zhang et al. 2015).

Stem cell therapy can be especially useful for treating blood diseases. In the hematopoietic system, self-renewing hematopoietic stem cells reside at the top of hierarchy, giving rise to various types of specialized blood cells (Seita and Weissman 2010). If healthy, long-term repopulating hematopoietic stem cells can be delivered to a patient's body, the production of healthy blood cells can last for the rest of the life for the patient. Particularly for FA, it has been known for long time that wild-type hematopoietic stem cells have a selective expansion and repopulating advantage over FA patient hematopoietic stem cells (Battaile et al. 1999). This will allow a small number of corrected hematopoietic stem cells to outgrow the patient's hematopoietic stem cells. It has been reported that only a few corrected hematopoietic stem cells can eventually repopulate the entire blood system in FA patients (Mankad et al. 2006; Waisfisz et al. 1999; Gross et al. 2002). Given these advantages of stem cell therapy, it is not surprising that allogeneic hematopoietic stem cell transplantation, the most commonly used stem cell therapy, is currently the only curative treatment to manage bone marrow failure for FA patients.

2 Allogeneic Hematopoietic Stem Cell Transplantation

Hematopoietic stem cells were the first tissue-specific stem cells to be isolated (Spangrude et al. 1988), and are so far the only type of stem cells used routinely in clinics to treat a variety of blood cell diseases (Bryder et al. 2006). After decades of practice, allogeneic hematopoietic stem cell transplantation (HSCT) has become a well-established treatment of many blood diseases, especially those that involve malfunctioning blood cells. In this process, multipotent hematopoietic stem cells are isolated from the bone marrow of an "immunologically matched" healthy donor and transplanted into the patient. Once inside of a patient's body, these hematopoietic stem cells colonize in the patient's bone marrow and give rise to all types of healthy mature blood cells.

In early days, donor hematopoietic cells were obtained by bone marrow harvest, a rather invasive procedure in which a large needle is used to remove bone marrow from the hipbone of the donor. Nowadays, most of allotransplants are obtained from mobilized peripheral blood. Treatment with mobilizing agents such as granulocyte colony-stimulating factor or plerixafor releases hematopoietic stem cells from the donor's bone marrow to peripheral blood (Lemoli and D'Addio 2008). Studies show that hematopoietic stem cell yield from mobilized peripheral blood is superior to that from a bone marrow harvest (Singhal et al. 2000).

For successful bone marrow transplantation, the key is to identify an "immunologically matched" healthy donor, often a sibling or a close relative. Human leukocyte antigens (HLA) are cell surface proteins responsible for the regulation of the immune system in humans and are used as markers to match a patient with a donor for transplantation. When a patient's HLA and the donor's HLA closely match, it is less likely that the patient's immune system will reject the donor cells. For patients without a perfect donor, partially mismatched donor can be used.

To facilitate the engraftment of the donor hematopoietic stem cells, the patient's bone marrow cells are usually destroyed by chemotherapy or radiation. The standard transplant conditioning regimens include fludarabine, cyclophosphamide, and total body irradiation. However, due to the underlying DNA repair deficiency in FA patients, these standard transplant conditioning regimens are particularly toxic to FA patients. Eliane Gluckman pioneered a reduced conditioning regimen (low-dose cyclophosphamide and thoracoabdominal irradiation) for FA patients and significantly improved the outcomes after allogeneic stem cell transplantation (Cavazzana 2014; Socie et al. 1998). This strategy has been further progressively optimized by other researchers in the field and significantly improved the survival rates of FA patients (Gluckman and Wagner 2008; MacMillan et al. 2009, 2011, 2015).

Currently, HSCT with HLA-identical sibling donors yields an excellent survival rate. Despite significant improvement in recent years, HSCT from HLA-mismatched donors is still challenging, with a higher risk of complications and lower survival rates.

Although allogeneic HSCT is the only cure available to most FA patients, this procedure carries significant risks. Because the recipient's immune system is destroyed before the transplantation, the patients have a higher risk for opportunistic infection. Also, graft failure occurs in as many as one third of the patients. In addition, one of the most common late complications after HSCT is graft-versus-host disease (GVHD), an illness in which the donor's immune cells attack the patient's own body and causes miserable side effects. About two thirds of all bone marrow recipients develop GVHD. For FA patients, the threat of GVHD can be more devastating: Due to the long-term immunosuppressive treatment for GVHD, HSCT patients have a higher risk of developing solid tumors such as oral squamous cell carcinoma (Demarosi et al. 2005; Kruse et al. 2009), a type of cancer to which FA patients are especially prone (Rosenberg et al. 2008).

3 Stem Cell Gene Therapy

To overcome the aforementioned complications and to find a cure for the FA patient who doesn't have a HLA-matched donor, a new therapy has emerged, in which autologous, gene-modified hematopoietic stem cells are used for transplantation. Instead of performing an allogeneic (the stem cells come from a donor) transplantation, the patient's own stem cells can be collected, corrected for the genetic defects, and returned to the patient via an autologous hematopoietic stem cell transplantation (Cavazzana 2014). This gene therapy approach will result in patient-specific genetic correction that fixes the underlying defect and thus provides an ultimate cure. Because the patient's own cells are used, this will eliminate the need for matched donors and the risk of graft-versus-host disease commonly associated with the conventional hematopoietic stem cell transplantation (Cavazzana 2014). A group of around 70 patients with either severe combined immunodeficiency conditions or adenosine deaminase deficiency have successfully undergone this type of gene therapy and validated this approach (Gaspar et al. 2011; Hacein-Bey-Abina et al. 2010).

Gene therapy is an experimental approach that modifies the expression of an individual's genes or that corrects abnormal genes. Gene therapy is a way to fix a genetic defect at its source. By adding a corrected copy of a defective gene, gene therapy restores the gene's function and helps diseased tissues and organs work properly. Most commonly, a healthy version of the gene is introduced into the body as an extra copy. This is called "gene addition" approach. Alternatively, the defected gene can be fixed by a "gene correction" approach through homologous recombination-mediated genome editing.

Generally speaking, genetic diseases caused by mutations in one gene are good candidates for gene therapy, while diseases associated with mutations in many genes or environmental factors are poor candidates. Gene therapy is particularly suitable for FA for a couple of other reasons. It has been repeatedly demonstrated that the defects of FA cells can be corrected by simply adding back a healthy copy of the defective gene. Genetic complementation has been shown to functionally correct the hematopoietic defects in FA mouse models and in FA patients with somatic mosaicism arising from spontaneous reversion of the genetic mutations (natural gene therapy) (Mankad et al. 2006; Waisfisz et al. 1999; Gross et al. 2002). Also, the 21 causative genes for the vast majority of FA patients are already cloned, and the disease-causing mutations are well-documented in Fanconi anemia mutation database (www.rockefeller.edu/fanconi/).

The first approved gene therapy in the United States took place in 1990. However, after a patient died from a massive immune response triggered by the use of the adenoviral vector in 1999, the Food and Drug Administration (FDA) suspended several clinical gene therapy trials. On the flipside, this incident has forced researchers to take a step back to more critically evaluate the safety and efficacy of a variety of gene therapy approaches. After decades of research, scientists can now design safer and more effective vectors, target different types of cells, and minimize deleterious immune responses in patients. Today, hundreds of gene therapy clinical trials are underway. Although the FDA hasn't approved any gene therapy products for commercial use in the United States yet, European regulators have approved Glybera for treating a pancreatic disorder and Strimvelis for treating children with genetic adenosine deaminase deficiency in recent years. Gene therapy is finally beginning to realize its long-sought promise.

Part of the challenge in gene therapy is to choose the most suitable vector for treating the disease of interest. Viral vectors are often used to carry altered genes into patients' cells. This is because they can easily enter cells, and they can also target specific types of cells. But sometimes they can cause immune responses in patients.

Also, in the case of retrovirus and lentivirus, the insertion of the genetic material can sometimes disrupt a tumor-suppressing gene or activate a pro-oncogene. This will potentially induce a tumor over time through a process called insertional oncogenesis. For this reason, non-viral vectors such as liposomes are also studied in clinical trials. Specifically for FA gene therapy, scientists have found that both retroviral (γ-retrovirus and foamy virus) and lentiviral vectors can successfully transduce FA patient cells and deliver a wild-type copy of the gene of interest into the cells (Becker et al. 2010; Galimi et al. 2002; Liu et al. 1997; Fu et al. 1997; Gush et al. 2000; Si et al. 2008).

There is one other challenge specific to FA patients. Clinical gene therapy targets CD34$^+$ cells, which are highly heterogeneous. The majority of these cells are multipotent progenitors and lineage-restricted cells, not hematopoietic stem cells. In order to achieve successful hematopoietic stem cell-based gene therapy, a sufficient amount of CD34$^+$ cells must be transduced to guarantee that hematopoietic stem cells are corrected. Prior clinical experience with autologous and allogeneic bone marrow transplantation suggests that a minimum of three million non-manipulated CD34$^+$ cells per kg body weight are necessary to restore hematopoiesis (Cavazzana 2014). However, patients with FA often have poor marrow cellularity. Thus, meeting this cell number requirement can be a major obstacle for gene therapy with FA patients (Kelly et al. 2007).

A potential solution is to expand hematopoietic stem cells from FA patients. In the past 20 years, numerous efforts have been made toward the ex vivo expansion of hematopoietic stem cells (Kiem et al. 2012). Among the many hematopoietic growth factors tested so far, the most successful expansion reagent identified is probably the aryl hydrocarbon receptor antagonist StemRegenin 1 (Kiem et al. 2012; Boitano et al. 2010). Additionally, some regents can specifically benefit FA cells, which are often more fragile in cell culture (Habi et al. 2005). For example, *N*-acetylcysteine has been proven to improve the survival of FA peripheral blood mononuclear cells in an ex vivo study (Monti et al. 1997).

So far there have been five registered clinical trials for gene therapy in FA (Table 1; also see details in clinicaltrials.gov). Three of them used retroviral vectors and had been completed. However, none of them generated clinical benefits related to the gene therapy procedures (Kelly et al. 2007; Liu et al. 1999). The international FA Gene Therapy Working Group has proposed the use of self-inactivating lentiviral vectors with phosphoglycerate kinase promoter and woodchuck hepatitis virus posttranscriptional regulatory element for a better balance between transgene expression and genotoxicity (Tolar et al. 2011, 2012). A proof-of-concept study using a similar strategy has demonstrated the efficacy of this design in a *Fanca* knockout mouse model (Molina-Estevez et al. 2015). The two ongoing FA gene therapy clinical trials (NCT01331018 and NCT02931071; see also Table 1) use a lentiviral vector with this design scheme to deliver a functional *Fanca* gene.

It is worth noting that all of these FA gene therapy clinical trials so far use the "gene addition" approach with viral vectors. Therefore, insertional oncogenesis can't be neglected before the safety of this approach is carefully established. Recent advances in genome engineering have expanded gene therapy options for

Table 1 Clinical trials of gene therapy for Fanconi anemia patients

Identifier	Genetic defect	Vector	Start	Status
NCT00001399	*Fancc*	Retrovirus	1999	Completed
NCT00005896	*Fancc*	Retrovirus	2000	Unknown
NCT00272857	*Fanca*	Retrovirus	2006	Completed
NCT01331018	*Fanca*	Lentivirus	2011	Recruiting
NCT02931071	*Fanca*	Lentivirus	2013	Recruiting

hematopoietic diseases. Unlike the "gene addition" strategy, "gene correction" approach uses a transient ex vivo intervention and does not cause permanent insertion of foreign DNA into the genome. A disadvantage of the "gene correction" approach, however, is its low efficiency because homologous recombination is a rare event. However, targeted nucleases, such as zinc-finger nucleases, transcription activator like effector nucleases (TALENs), and CRISPR (clustered, regularly interspaced, short palindromic repeats)-associated RNA-guided nuclease Cas9, can induce site-specific double-strand breaks into the human genome and boost homologous recombination by up to 1000-fold (Hockemeyer et al. 2009, 2011; Zou et al. 2009).

As a first step toward FA gene correction using CRISPR/Cas9 system (Hsu et al. 2014), researchers have recently achieved precise genome editing of cultured FA patient-derived fibroblasts in vitro (Osborn et al. 2015). Moving forward, the challenge is how to efficiently deliver these targeted nucleases in vivo while avoiding the viral vectors associated with potential risk for insertional mutagenesis. One of the options is nanoparticles, which have recently successfully employed to deliver Cas9 in a mouse model of genetic liver disease (Yin et al. 2016). Alternatively, these nucleases can be delivered by the newer generation, nonintegrating adeno-associated virus (AAV) vectors. AAV is a particularly attractive tool for gene transfer, as it is non-pathogenic; can infect a broad host range of cells, including non-dividing cells; and rarely integrates into the genome. Compared with retroviral or lentiviral vectors, AAV carries low immunogenic potential and reduced oncogenic risk from host genome integration. Nonetheless, further research is warranted to explore these state-of-art gene-editing tools for FA stem cell gene therapy.

4 Alternative Sources of Hematopoietic Stem Cells

Donor stem cell material for hematopoietic transplantation is a limited resource. For the patients who lack an HLA-matched unrelated bone marrow donor, alternative sources of transplantable allogeneic or patient-specific hematopoietic stem cells are badly needed. In recent years, advances have been made to develop stem cells from various sources, including umbilical cord blood and induced pluripotent stem cells.

Umbilical cord blood is rich in hematopoietic stem cells, similar to those found in bone marrow. Therefore, cord blood donated to public blood banks provides another source of hematopoietic stem cells. Cord blood is being used increasingly on an experimental basis for stem cell transplantation as an alternative resource to bone marrow. Cord blood collection is relatively effortless and carries no health risk to the donors. Also, studies have shown that, unlike the traditional bone marrow transplantation, cord blood transplantation only requires the donor and the recipient to be partially matched. In recent years, stem cell therapy using cord blood-derived hematopoietic stem cells has been used to cure blood and immunological diseases (Wagner and Gluckman 2010). A subset of FA patients has been treated with cord blood transplantation (Gluckman et al. 2007). A formal comparison of patient outcomes between this strategy and the traditional bone marrow transplantation is not available yet. In general, unrelated cord blood transplantation has a low incidence of GVHD and may be advantageous for FA patients. However, the progress in expanding cord blood usage is very slow (Rosemann 2014). The main problem is that the quantity of hematopoietic stem cells from cord blood is much smaller than that in peripheral blood or bone marrow, due to the small volume of collected cord blood. As a consequence, bone marrow engraftment is slower with cord blood transplantation. Patients, therefore, may be more vulnerable to opportunistic infection. This problem is more pronounced for adolescent and adult recipients, as they often need more stem cells for transplantation.

Another resource is pluripotent stem cells. Induced pluripotent stem cells (iPS) were first described in 2006 and soon showed immense promise (Takahashi and Yamanaka 2006). By overexpressing four factors (Oct4, Sox2, c-Myc, and Klf4), Takahashi and Yamanaka reprogramed specialized somatic cells, such as skin cells, to functionally pluripotent stem cells. Using this technique, somatic cells from a patient with a genetic disease can be reprogrammed to iPS cells, genetically corrected, and used for regenerative medicine. This approach holds the prospect of generating an abundant supply of customized, patient-specific pluripotent stem cells that are immune-compatible for stem cell therapy.

It has been discovered that the DNA repair defects in FA cells hinder the reprogramming process and make these cells relatively resistant to reprogramming (Raya et al. 2009). To overcome this obstacle, FA patient cells can be first corrected for the genetic defect before reprogramming (Raya et al. 2009; Muller et al. 2012). Further efforts are also being made to improve the current reprogramming methods (Muller et al. 2012).

Patient-specific pluripotent stem cells can be turned to hematopoietic stem cells and used for autologous transplantation to treat a blood disease (Robinton and Daley 2012). In a proof-of-concept experiment, researchers used gene-targeting technique to correct the genetic defect in iPS cells derived from a humanized sickle cell anemia mouse model. They then differentiated these repaired iPS cells into hematopoietic progenitors and delivered them back to the mice. This approach successfully corrected the genetic disease in the mice (Hanna et al. 2007).

One obstacle remains before this approach can be generally applied to the clinic. Despite some progress, the attempts to turn human iPS cells to functional mature HSPCs have been largely unsuccessful (Daniel et al. 2016). Encouragingly, there has been one successful attempt to re-differentiate corrected FA patient-derived iPS cells into blood progenitors in vitro (Raya et al. 2009).

Overall, the use of reprogrammed pluripotent stem cells for regenerative medicine is still being tested in experimental labs. There is no indication that the patient-specific cell reprogramming approach will become a viable option for autologous hematopoietic stem cell transplantation in the near future. However, once the problems described above are solved, this strategy holds an unlimited therapeutic potential.

5 Conclusions

Allogeneic hematopoietic stem cell transplantation is currently the only curative treatment to manage bone marrow failure for FA patients. Despite recent advances that have significantly improved FA patients' survival rate, this procedure still has its limitations and complications. But FA patients can benefit more from treatments that combine the technologies of gene and cell therapy. In recent years, new developments in gene therapy may finally make autologous, gene-modified stem cell therapy a viable option for patients with FA and other blood diseases. Numerous ongoing clinical trials will establish the safety and efficacy of this approach. The rapidly evolving gene engineering tools will greatly facilitate achieving this goal in the coming years. In addition, patient-specific pluripotent stem cells might be harnessed to generate hematopoietic stem cells and expand the currently limited source material available for transplantation.

References

Auerbach, A. D. (2009). Fanconi anemia and its diagnosis. *Mutation Research, 668*(1–2), 4–10. doi:10.1016/j.mrfmmm.2009.01.013.

Battaile, K. P., Bateman, R. L., Mortimer, D., Mulcahy, J., Rathbun, R. K., Bagby, G., Fleming, W. H., & Grompe, M. (1999). In vivo selection of wild-type hematopoietic stem cells in a murine model of Fanconi anemia. *Blood, 94*(6), 2151–2158.

Becker, P. S., Taylor, J. A., Trobridge, G. D., Zhao, X., Beard, B. C., Chien, S., Adair, J., Kohn, D. B., Wagner, J. E., Shimamura, A., & Kiem, H. P. (2010). Preclinical correction of human Fanconi anemia complementation group a bone marrow cells using a safety-modified lentiviral vector. *Gene Therapy, 17*(10), 1244–1252. doi:10.1038/gt.2010.62.

Boitano, A. E., Wang, J., Romeo, R., Bouchez, L. C., Parker, A. E., Sutton, S. E., Walker, J. R., Flaveny, C. A., Perdew, G. H., Denison, M. S., Schultz, P. G., &

Cooke, M. P. (2010). Aryl hydrocarbon receptor antagonists promote the expansion of human hematopoietic stem cells. *Science, 329*(5997), 1345–1348. doi:10.1126/science.1191536.

Bryder, D., Rossi, D. J., & Weissman, I. L. (2006). Hematopoietic stem cells: The paradigmatic tissue-specific stem cell. *The American Journal of Pathology, 169*(2), 338–346. doi:10.2353/ajpath.2006.060312.

Cavazzana, M. (2014). Hematopoietic stem cell gene therapy: Progress on the clinical front. *Human Gene Therapy, 25*(3), 165–170. doi:10.1089/hum.2014.2504.

Ceccaldi, R., Briot, D., Larghero, J., Vasquez, N., Dubois d'Enghien, C., Chamousset, D., Noguera, M. E., Waisfisz, Q., Hermine, O., Pondarre, C., Leblanc, T., Gluckman, E., Joenje, H., Stoppa-Lyonnet, D., Socie, G., & Soulier, J. (2011). Spontaneous abrogation of the G(2)DNA damage checkpoint has clinical benefits but promotes leukemogenesis in Fanconi anemia patients. *The Journal of Clinical Investigation, 121* (1), 184–194. doi:10.1172/JCI43836.

Ceccaldi, R., Parmar, K., Mouly, E., Delord, M., Kim, J. M., Regairaz, M., Pla, M., Vasquez, N., Zhang, Q. S., Pondarre, C., Peffault de Latour, R., Gluckman, E., Cavazzana-Calvo, M., Leblanc, T., Larghero, J., Grompe, M., Socie, G., D'Andrea, A. D., & Soulier, J. (2012). Bone marrow failure in Fanconi anemia is triggered by an exacerbated p53/p21 DNA damage response that impairs hematopoietic stem and progenitor cells. *Cell Stem Cell, 11*(1), 36–49. doi:10.1016/j.stem.2012.05.013.

Daniel, M. G., Pereira, C. F., Lemischka, I. R., & Moore, K. A. (2016). Making a hematopoietic stem cell. *Trends in Cell Biology, 26*(3), 202–214. doi:10.1016/j.tcb.2015.10.002.

Demarosi, F., Lodi, G., Carrassi, A., Soligo, D., & Sardella, A. (2005). Oral malignancies following HSCT: Graft versus host disease and other risk factors. *Oral Oncology, 41*(9), 865–877. doi:10.1016/j.oraloncology.2005.02.001.

Dokal, I. (2003). Inherited aplastic anaemia. *The Hematology Journal: The Official Journal of the European Haematology Association/EHA, 4*(1), 3–9. doi:10.1038/sj.thj.6200215.

Dong, H., Nebert, D. W., Bruford, E. A., Thompson, D. C., Joenje, H., & Vasiliou, V. (2015). Update of the human and mouse Fanconi anemia genes. *Human Genomics, 9*(1), 32. doi:10.1186/s40246-015-0054-y.

Fu, K. L., Foe, J. R., Joenje, H., Rao, K. W., Liu, J. M., & Walsh, C. E. (1997). Functional correction of Fanconi anemia group A hematopoietic cells by retroviral gene transfer. *Blood, 90*(9), 3296–3303.

Galimi, F., Noll, M., Kanazawa, Y., Lax, T., Chen, C., Grompe, M., & Verma, I. M. (2002). Gene therapy of Fanconi anemia: Preclinical efficacy using lentiviral vectors. *Blood, 100*(8), 2732–2736. doi:10.1182/blood-2002-04-1245.

Gaspar, H. B., Cooray, S., Gilmour, K. C., Parsley, K. L., Zhang, F., Adams, S., Bjorkegren, E., Bayford, J., Brown, L., Davies, E. G., Veys, P., Fairbanks, L., Bordon, V., Petropoulou, T., Kinnon, C., & Thrasher, A. J. (2011). Hematopoietic stem cell gene therapy for adenosine deaminase-deficient severe combined immunodeficiency leads to long-term immunological recovery and metabolic correction. *Science Translational Medicine, 3*(97), 97ra80. doi:10.1126/scitranslmed.3002716.

Gluckman, E., & Wagner, J. E. (2008). Hematopoietic stem cell transplantation in childhood inherited bone marrow failure syndrome. *Bone Marrow Transplantation, 41*(2), 127–132. doi:10.1038/sj.bmt.1705960.

Gluckman, E., Rocha, V., Ionescu, I., Bierings, M., Harris, R. E., Wagner, J., Kurtzberg, J., Champagne, M. A., Bonfim, C., Bittencourt, M., Darbyshire, P., Fernandez, M. N., Locatelli, F., & Pasquini, R. (2007). Results of unrelated cord blood transplant in Fanconi anemia patients: Risk factor analysis for engraftment and survival. *Biology of Blood and Marrow Transplantation: Journal of the American Society for Blood and Marrow Transplantation, 13*(9), 1073–1082. doi:10.1016/j.bbmt.2007.05.015.

Gross, M., Hanenberg, H., Lobitz, S., Friedl, R., Herterich, S., Dietrich, R., Gruhn, B., Schindler, D., & Hoehn, H. (2002). Reverse mosaicism in Fanconi anemia: Natural gene therapy via molecular self-correction. *Cytogenet Genome Research, 98*(2–3), 126–135. doi:69805.

Gush, K. A., Fu, K. L., Grompe, M., & Walsh, C. E. (2000). Phenotypic correction of Fanconi anemia group C knockout mice. *Blood, 95*(2), 700–704.

Habi, O., Delisle, M. C., Messier, N., & Carreau, M. (2005). Lack of self-renewal capacity in Fancc−/− stem cells after ex vivo expansion. *Stem Cells, 23* (8), 1135–1141. doi:10.1634/stemcells.2004-0356.

Hacein-Bey-Abina, S., Hauer, J., Lim, A., Picard, C., Wang, G. P., Berry, C. C., Martinache, C., Rieux-Laucat, F., Latour, S., Belohradsky, B. H., Leiva, L., Sorensen, R., Debre, M., Casanova, J. L., Blanche, S., Durandy, A., Bushman, F. D., Fischer, A., & Cavazzana-Calvo, M. (2010). Efficacy of gene therapy for X-linked severe combined immunodeficiency. *The New England Journal of Medicine, 363*(4), 355–364. doi:10.1056/NEJMoa1000164.

Hanna, J., Wernig, M., Markoulaki, S., Sun, C. W., Meissner, A., Cassady, J. P., Beard, C., Brambrink, T., Wu, L. C., Townes, T. M., & Jaenisch, R. (2007). Treatment of sickle cell anemia mouse model with iPS cells generated from autologous skin. *Science, 318* (5858), 1920–1923. doi:10.1126/science.1152092.

Hockemeyer, D., Soldner, F., Beard, C., Gao, Q., Mitalipova, M., DeKelver, R. C., Katibah, G. E., Amora, R., Boydston, E. A., Zeitler, B., Meng, X., Miller, J. C., Zhang, L., Rebar, E. J., Gregory, P. D., Urnov, F. D., & Jaenisch, R. (2009). Efficient targeting of expressed and silent genes in human ESCs and iPSCs using zinc-finger nucleases. *Nature*

Biotechnology, 27(9), 851–857. doi:10.1038/nbt.1562.

Hockemeyer, D., Wang, H., Kiani, S., Lai, C. S., Gao, Q., Cassady, J. P., Cost, G. J., Zhang, L., Santiago, Y., Miller, J. C., Zeitler, B., Cherone, J. M., Meng, X., Hinkley, S. J., Rebar, E. J., Gregory, P. D., Urnov, F. D., & Jaenisch, R. (2011). Genetic engineering of human pluripotent cells using TALE nucleases. *Nature Biotechnology, 29*(8), 731–734. doi:10.1038/nbt.1927.

Hsu, P. D., Lander, E. S., & Zhang, F. (2014). Development and applications of CRISPR-Cas9 for genome engineering. *Cell, 157*(6), 1262–1278. doi:10.1016/j.cell.2014.05.010.

Kelly, P. F., Radtke, S., von Kalle, C., Balcik, B., Bohn, K., Mueller, R., Schuesler, T., Haren, M., Reeves, L., Cancelas, J. A., Leemhuis, T., Harris, R., Auerbach, A. D., Smith, F. O., Davies, S. M., & Williams, D. A. (2007). Stem cell collection and gene transfer in Fanconi anemia. *Molecular Therapy, 15*(1), 211–219. doi:10.1038/sj.mt.6300033.

Kiem, H. P., Jerome, K. R., Deeks, S. G., & McCune, J. M. (2012). Hematopoietic-stem-cell-based gene therapy for HIV disease. *Cell Stem Cell, 10*(2), 137–147. doi:10.1016/j.stem.2011.12.015.

Kim, H., & D'Andrea, A. D. (2012). Regulation of DNA cross-link repair by the Fanconi anemia/BRCA pathway. *Genes & Development, 26*(13), 1393–1408. doi:10.1101/gad.195248.112.

Kruse, A. L., Zwahlen, R. A., & Gratz, K. W. (2009). New classification of maxillary ameloblastic carcinoma based on an evidence-based literature review over the last 60 years. *Head & Neck Oncology, 1*, 31. doi:10.1186/1758-3284-1-31.

Kutler, D. I., Singh, B., Satagopan, J., Batish, S. D., Berwick, M., Giampietro, P. F., Hanenberg, H., & Auerbach, A. D. (2003). A 20-year perspective on the International Fanconi Anemia Registry (IFAR). *Blood, 101*(4), 1249–1256. doi:10.1182/blood-2002-07-2170.

Lemoli, R. M., & D'Addio, A. (2008). Hematopoietic stem cell mobilization. *Haematologica, 93*(3), 321–324. doi:10.3324/haematol.12616.

Liu, J. M., Young, N. S., Walsh, C. E., Cottler-Fox, M., Carter, C., Dunbar, C., Barrett, A. J., & Emmons, R. (1997). Retroviral mediated gene transfer of the Fanconi anemia complementation group C gene to hematopoietic progenitors of group C patients. *Human Gene Therapy, 8*(14), 1715–1730. doi:10.1089/hum.1997.8.14-1715.

Liu, J. M., Kim, S., Read, E. J., Futaki, M., Dokal, I., Carter, C. S., Leitman, S. F., Pensiero, M., Young, N. S., & Walsh, C. E. (1999). Engraftment of hematopoietic progenitor cells transduced with the Fanconi anemia group C gene (FANCC). *Human Gene Therapy, 10*(14), 2337–2346. doi:10.1089/10430349950016988.

Macmillan, M. L., Blazar, B. R., DeFor, T. E., & Wagner, J. E. (2009). Transplantation of ex-vivo culture-expanded parental haploidentical mesenchymal stem cells to promote engraftment in pediatric recipients of unrelated donor umbilical cord blood: Results of a phase I-II clinical trial. *Bone Marrow Transplantation, 43*(6), 447–454. doi:10.1038/bmt.2008.348.

MacMillan, M. L., Hughes, M. R., Agarwal, S., & Daley, G. Q. (2011). Cellular therapy for fanconi anemia: The past, present, and future. *Biology of Blood and Marrow Transplantation: Journal of the American Society for Blood and Marrow Transplantation, 17*(1 Suppl), S109–S114. doi:10.1016/j.bbmt.2010.11.027.

MacMillan, M. L., DeFor, T. E., Young, J. A., Dusenbery, K. E., Blazar, B. R., Slungaard, A., Zierhut, H., Weisdorf, D. J., & Wagner, J. E. (2015). Alternative donor hematopoietic cell transplantation for Fanconi anemia. *Blood, 125*(24), 3798–3804. doi:10.1182/blood-2015-02-626002.

Mankad, A., Taniguchi, T., Cox, B., Akkari, Y., Rathbun, R. K., Lucas, L., Bagby, G., Olson, S., D'Andrea, A., & Grompe, M. (2006). Natural gene therapy in monozygotic twins with Fanconi anemia. *Blood, 107*(8), 3084–3090. doi:10.1182/blood-2005-07-2638.

Molina-Estevez, F. J., Nowrouzi, A., Lozano, M. L., Galy, A., Charrier, S., von Kalle, C., Guenechea, G., Bueren, J. A., & Schmidt, M. (2015). Lentiviral-mediated Gene therapy in Fanconi anemia-A mice reveals long-term engraftment and continuous turnover of corrected HSCs. *Current Gene Therapy, 15*(6), 550–562.

Monti, D., Macchioni, S., Guido, M., Pagano, G., Zatterale, A., Calzone, R., Cossarizza, A., Straface, E., Malorni, W., & Franceschi, C. (1997). Resistance to apoptosis in Fanconi's anaemia. An ex vivo study in peripheral blood mononuclear cells. *FEBS Letters, 409*(3), 365–369.

Muller, L. U., Milsom, M. D., Harris, C. E., Vyas, R., Brumme, K. M., Parmar, K., Moreau, L. A., Schambach, A., Park, I. H., London, W. B., Strait, K., Schlaeger, T., Devine, A. L., Grassman, E., D'Andrea, A., Daley, G. Q., & Williams, D. A. (2012). Overcoming reprogramming resistance of Fanconi anemia cells. *Blood, 119*(23), 5449–5457. doi:10.1182/blood-2012-02-408674.

Osborn, M. J., Gabriel, R., Webber, B. R., DeFeo, A. P., McElroy, A. N., Jarjour, J., Starker, C. G., Wagner, J. E., Joung, J. K., Voytas, D. F., von Kalle, C., Schmidt, M., Blazar, B. R., & Tolar, J. (2015). Fanconi anemia gene editing by the CRISPR/Cas9 system. *Human Gene Therapy, 26*(2), 114–126. doi:10.1089/hum.2014.111.

Raya, A., Rodriguez-Piza, I., Guenechea, G., Vassena, R., Navarro, S., Barrero, M. J., Consiglio, A., Castella, M., Rio, P., Sleep, E., Gonzalez, F., Tiscornia, G., Garreta, E., Aasen, T., Veiga, A., Verma, I. M., Surralles, J., Bueren, J., & Izpisua Belmonte, J. C. (2009). Disease-corrected haematopoietic progenitors from Fanconi anaemia induced pluripotent stem cells. *Nature, 460*(7251), 53–59. doi:10.1038/nature08129.

Robinton, D. A., & Daley, G. Q. (2012). The promise of induced pluripotent stem cells in research and therapy. *Nature, 481*(7381), 295–305. doi:10.1038/nature10761.

Rosemann, A. (2014). Why regenerative stem cell medicine progresses slower than expected. *Journal of Cellular Biochemistry, 115*(12), 2073–2076. doi:10.1002/jcb.24894.

Rosenberg, P. S., Alter, B. P., & Ebell, W. (2008). Cancer risks in Fanconi anemia: Findings from the German Fanconi anemia registry. *Haematologica, 93*(4), 511–517. doi:10.3324/haematol.12234.

Seita, J., & Weissman, I. L. (2010). Hematopoietic stem cell: Self-renewal versus differentiation. *Wiley Interdisciplinary Reviews Systems Biology and Medicine, 2*(6), 640–653. doi:10.1002/wsbm.86.

Shimamura, A., & Alter, B. P. (2010). Pathophysiology and management of inherited bone marrow failure syndromes. *Blood Reviews, 24*(3), 101–122. doi:10.1016/j.blre.2010.03.002.

Si, Y., Pulliam, A. C., Linka, Y., Ciccone, S., Leurs, C., Yuan, J., Eckermann, O., Fruehauf, S., Mooney, S., Hanenberg, H., & Clapp, D. W. (2008). Overnight transduction with foamyviral vectors restores the long-term repopulating activity of Fancc−/− stem cells. *Blood, 112*(12), 4458–4465. doi:10.1182/blood-2007-07-102947.

Singhal, S., Powles, R., Kulkarni, S., Treleaven, J., Sirohi, B., Millar, B., Shepherd, V., Saso, R., Rowland, A., Long, S., Cabral, S., Horton, C., & Mehta, J. (2000). Comparison of marrow and blood cell yields from the same donors in a double-blind, randomized study of allogeneic marrow vs blood stem cell transplantation. *Bone Marrow Transplantation, 25*(5), 501–505. doi:10.1038/sj.bmt.1702173.

Socie, G., Devergie, A., Girinski, T., Piel, G., Ribaud, P., Esperou, H., Parquet, N., Maarek, O., Noguera, M. H., Richard, P., Brison, O., & Gluckman, E. (1998). Transplantation for Fanconi's anaemia: Long-term follow-up of fifty patients transplanted from a sibling donor after low-dose cyclophosphamide and thoraco-abdominal irradiation for conditioning. *British Journal of Haematology, 103*(1), 249–255.

Spangrude, G. J., Heimfeld, S., & Weissman, I. L. (1988). Purification and characterization of mouse hematopoietic stem cells. *Science, 241*(4861), 58–62.

Takahashi, K., & Yamanaka, S. (2006). Induction of pluripotent stem cells from mouse embryonic and adult fibroblast cultures by defined factors. *Cell, 126*(4), 663–676. doi:10.1016/j.cell.2006.07.024.

Tolar, J., Adair, J. E., Antoniou, M., Bartholomae, C. C., Becker, P. S., Blazar, B. R., Bueren, J., Carroll, T., Cavazzana-Calvo, M., Clapp, D. W., Dalgleish, R., Galy, A., Gaspar, H. B., Hanenberg, H., Von Kalle, C., Kiem, H. P., Lindeman, D., Naldini, L., Navarro, S., Renella, R., Rio, P., Sevilla, J., Schmidt, M., Verhoeyen, E., Wagner, J. E., Williams, D. A., & Thrasher, A. J. (2011). Stem cell gene therapy for Fanconi anemia: Report from the 1st international Fanconi anemia gene therapy working group meeting. *Molecular Therapy, 19*(7), 1193–1198. doi:10.1038/mt.2011.78.

Tolar, J., Becker, P. S., Clapp, D. W., Hanenberg, H., de Heredia, C. D., Kiem, H. P., Navarro, S., Qasba, P., Rio, P., Schmidt, M., Sevilla, J., Verhoeyen, E., Thrasher, A. J., & Bueren, J. (2012). Gene therapy for Fanconi anemia: One step closer to the clinic. *Human Gene Therapy, 23*(2), 141–144. doi:10.1089/hum.2011.237.

Wagner, J. E., & Gluckman, E. (2010). Umbilical cord blood transplantation: The first 20 years. *Seminars in Hematology, 47*(1), 3–12. doi:10.1053/j.seminhematol.2009.10.011.

Waisfisz, Q., Morgan, N. V., Savino, M., de Winter, J. P., van Berkel, C. G., Hoatlin, M. E., Ianzano, L., Gibson, R. A., Arwert, F., Savoia, A., Mathew, C. G., Pronk, J. C., & Joenje, H. (1999). Spontaneous functional correction of homozygous Fanconi anaemia alleles reveals novel mechanistic basis for reverse mosaicism. *Nature Genetics, 22*(4), 379–383. doi:10.1038/11956.

Yin, H., Song, C. Q., Dorkin, J. R., Zhu, L. J., Li, Y., Wu, Q., Park, A., Yang, J., Suresh, S., Bizhanova, A., Gupta, A., Bolukbasi, M. F., Walsh, S., Bogorad, R. L., Gao, G., Weng, Z., Dong, Y., Koteliansky, V., Wolfe, S. A., Langer, R., Xue, W., & Anderson, D. G. (2016). Therapeutic genome editing by combined viral and non-viral delivery of CRISPR system components in vivo. *Nature Biotechnology*. doi:10.1038/nbt.3471.

Zhang, Q. S., Benedetti, E., Deater, M., Schubert, K., Major, A., Pelz, C., Impey, S., Marquez-Loza, L., Rathbun, R. K., Kato, S., Bagby, G. C., & Grompe, M. (2015). Oxymetholone therapy of fanconi anemia suppresses osteopontin transcription and induces hematopoietic stem cell cycling. *Stem Cell Reports, 4*(1), 90–102. doi:10.1016/j.stemcr.2014.10.014.

Zou, J., Maeder, M. L., Mali, P., Pruett-Miller, S. M., Thibodeau-Beganny, S., Chou, B. K., Chen, G., Ye, Z., Park, I. H., Daley, G. Q., Porteus, M. H., Joung, J. K., & Cheng, L. (2009). Gene targeting of a disease-related gene in human induced pluripotent stem and embryonic stem cells. *Cell Stem Cell, 5*(1), 97–110. doi:10.1016/j.stem.2009.05.023.

Advs Exp. Medicine, Biology - Innovations in Cancer Research and Regenerative Medicine (2018) 1: 29–44
DOI 10.1007/5584_2017_74

Published online: 21 July 2017

Pooled Human Serum Increases Regenerative Potential of In Vitro Expanded Stem Cells from Human Extracted Deciduous Teeth

Nazmul Haque and Noor Hayaty Abu Kasim

Abstract

In regenerative therapy, in vitro expansion of stem cells is critical to obtain a significantly higher number of cells for successful engraftment after transplantation. However, stem cells lose its regenerative potential and enter senescence during in vitro expansion. In this study, the influence of foetal bovine serum (FBS) and pooled human serum (pHS) on the proliferation, morphology and migration of stem cells from human extracted deciduous teeth (SHED) was compared. SHED ($n = 3$) was expanded in KnockOut DMEM supplemented with either pHS (pHS-SM) or FBS (FBS-SM). pHS was prepared using peripheral blood serum of six healthy male adults, aged between 21 and 35 years old. The number of live SHED was significantly higher, from passage 5 to 7, when cultured in pHS-SM compared to those cultured in FBS-SM ($p < 0.05$). Number of cells having flattened morphology, characteristics of partially differentiated and senescent cells, was significantly lower ($p < 0.05$) in pHS-SM (3%) compared to those in FBS-SM (7%). Furthermore, migration of SHED in pHS-SM was found to be more directional. The presence of selected ten paracrine factors known for their proliferation and migration potential was detected in all six individual human sera, used to produce pHS, none of which were detected in FBS. Ingenuity Pathway Analysis showed the possible involvement of the 'ephrin receptor signalling pathway' to regulate the proliferation and migration of SHED in

N. Haque and N.H. Abu Kasim (✉)
Department of Restorative Dentistry, Faculty of Dentistry, University of Malaya, Kuala Lumpur 50603, Malaysia

Regenerative Dentistry Research Group, Faculty of Dentistry, University of Malaya, Kuala Lumpur, Malaysia
e-mail: tanna.bge@gmail.com; nhayaty@um.edu.my

pHS-SM. In conclusion, pHS-SM showed significantly higher proliferation rate and could maintain significantly lower number of senescent cells and support directional migration of cells.

Keywords
Engraftment • Foetal bovine serum • Morphology • Paracrine factors • Regenerative medicine

Abbreviations

BM	bone marrow
EGF	epidermal growth factor
FBS	foetal bovine serum
FC	flattened
FGF-2	fibroblast growth factor-2
FSS	flat spindle-shaped
G-CSF	granulocyte colony-stimulating factor
GM-CSF	granulocyte macrophage colony-stimulating factor
HGF	hepatocyte growth factor
LIF	leukaemia inhibitory factor
MSCs	mesenchymal stem cells
PDGF-BB	platelet-derived growth factor BB
pHS	pooled human serum
RS	rapidly self-renewing
SCF	stem cell factor
SDF-1α	stromal cell-derived factor-1α
SHED	stem cells from human extracted deciduous teeth
SS	spindle-shaped
VEGF	vascular endothelial growth factor

1 Introduction

In vitro expansion of mesenchymal stem cells (MSCs) is critical to yield adequate number of cells (200–400 million) needed for transplantation for each regenerative therapy (Pittenger et al. 1999). During in vitro expansion, MSCs are likely to enter senescence (Bonab et al. 2006; Estrada et al. 2012) and lose migration potential (Honczarenko et al. 2006), which are considered as the major causes for poor engraftment upon transplantation, while rapid proliferation potential necessary to maintain the stemness (Saller et al. 2012; Lee et al. 2006) and site-specific migration and homing are vital for the successful engraftment (Eggenhofer et al. 2014).

On the basis of morphology and proliferation potential, MSCs are divided into the four groups, namely, rapidly self-renewing (RS), flat spindle-shaped (FSS), spindle-shaped (SS) and flattened (FC) (Haasters et al. 2009; Saller et al. 2012; Colter et al. 2001). Use of in vitro expanded MSCs with a higher number of RSs having the highest self-renewal potential and stemness (Lee et al. 2006; Haasters et al. 2009) and lower number of FCs having the lowest self-renewable capacity (Colter et al. 2001) might be more efficient in engraftment.

A number of paracrine factors, namely, vascular endothelial growth factor (VEGF), platelet-derived growth factor (PDGF) BB, fibroblast growth factor-2 (FGF-2), hepatocyte growth factor (HGF) and epidermal growth factor (EGF), can boost MSC proliferation (Rodrigues et al. 2010). However, use of these paracrine factors during in vitro expansion may cause priming and influence differentiation potential of MSCs (Handorf and Li 2011; Prasanna et al. 2010). During in vitro expansion, MSCs lose their site-specific migration potential (Honczarenko et al. 2006) consequently the regenerative outcomes. Hence, the combination of paracrine factors in the serum supplement during in vitro expansion determines the stemness as well as the proliferation and migration potential of MSCs. Earlier it was reported that pooled human serum (pHS) from adult AB blood donors and pooled cord blood serum maintain the differentiation potential, motility and immunosuppressive properties of MSCs (Kobayashi et al. 2005; Phadnis et al. 2006; Poloni et al. 2009; Tateishi et al. 2008; Eubanks et al. 2014; Li et al. 2014). However, similar studies on pHS using blood from donors irrespective of their blood groups are yet to be conducted. Again, the composition of the paracrine factors of xenogeneic serum such as foetal bovine serum (FBS) varies from that in the HS. Hence, it is expected that MSCs might respond differently when cultured in pHS-supplemented media (pHS-SM) compared to the FBS-supplemented media (FBS-SM), especially when targeted for in vitro expansion for regenerative therapy. Therefore, in this study we have compared the influence of pHS and FBS on the proliferation, morphology and migration of stem cells from human extracted deciduous teeth (SHED). Indeed, SHED have been investigated as a suitable source of MSCs for potential regenerative therapy because of their easy accessibility, pose no risk to the donor and have competitive self-renewal capability (Wang et al. 2012; Nakamura et al. 2009).

2 Materials and Methods

2.1 Ethics Approval for the Collection of SHED and Blood

Sample collection procedure for the current research was approved by the Medical Ethics Committee, Faculty of Dentistry, University of Malaya (reference: DFRD1301/0012[L] for blood collection; DFCO1107/0066[L] for teeth collection). Samples were collected following written consent from the donors or their guardians.

2.2 Isolation and Expansion of SHED

Sound intact deciduous molars were extracted from children ($n = 3$; aged 5–9 years) who were undergoing a planned serial extraction for management of occlusion at the Department of Paediatric Dentistry and Orthodontics, Faculty of Dentistry, University of Malaya. SHED were isolated according to the established procedure described before (Govindasamy et al. 2010).

In brief, following extraction, the root surfaces of teeth were cleaned with povidone-iodine (Sigma-Aldrich, St Louis, MO, USA), and the teeth were then placed into sterile solution prior to sectioning. Teeth were sectioned at the cement-enamel junction using a diamond rotary disc, and the dental pulp was removed with a sterile endodontic broach and kept in transportation medium. Following three times washing with washing buffer prepared by mixing equal volume of DPBS (Gibco, Grand Island, NY, USA) and penicillin-streptomycin (Gibco), the dental pulp was minced by using sterile scalpel followed by enzymatic digestion in 1% (w/v) collagenase type I (Gibco) for 40 min at 37 °C. After that collagenase was neutralized using medium containing 10% FBS (Gibco) and was centrifuged at 1250 rpm for 6 min. Following centrifugation pellet was resuspended in KnockOut™ DMEM (Gibco) with 10% FBS (Gibco), 1% GlutaMAX (Gibco) and 0.5% penicillin-streptomycin antibiotics (Gibco) and seeded in T25 flasks (BD Bioscience, Franklin Lakes, NJ, USA). Flasks were left at 37 °C and 5% CO_2 in humidified chambers.

Between day 3 and day 5 of incubation, growth of stem cells from primary tissue was seen under inverted microscope (Primo Vert, Carl Zeiss, Jena, Germany). On day 14, the first subculture of the SHED from primary dental pulp tissues was done.

2.3 Identification of MSC

According to the International Society for Cellular Therapy (ISCT), MSCs should have adherence to plastic, specific surface antigen (Ag) expression and multipotent differentiation potential (Dominici et al. 2006).

Plastic adherence of SHED was confirmed by viewing the culture flask under inverted microscope (Primo Vert, Carl Zeiss). Expression of specific surface antigen on SHED was determined by staining cells with human MSC phenotyping kit (Miltenyi Biotec, Bergisch Gladbach, Germany) and using MACSQuant® Analyzer 10 flow cytometer (Miltenyi Biotec). In brief, after reaching 70% confluency, SHED were dissociated using TrypLE™ Express (Gibco). A number of nucleated cells were counted using trypan blue (Gibco) and aliquoted (1×10^6 cells each) into seven pre-labelled (i.e. 1, 2, PerCP, PE, APC, FITC and blank) microcentrifuge tubes. Cells were centrifuged at 300 × g for 10 min and supernatants were aspirated completely. Cells in the tube labelled 'blank' were resuspended into 500 μl of buffer (2% FBS containing DPBS) and kept in a container containing ice. Cells in the other six tubes were resuspended into 100 μl buffer. After that, 10 μl of the MSC phenotyping cocktail (antihuman: CD14-PerCP, CD20-PerCP, CD34-PerCP, CD45-PerCP, CD73-APC, CD90-FITC, CD105-PE), 10 μl of the isotype control cocktail (antihuman: mouse IgG1-FITC, mouse IgG1-PE, mouse IgG1-APC, mouse IgG1-PerCP, mouse IgG2a-PerCP), 10 μl of antihuman CD73-Biotin, 10 μl of antihuman CD105-PE, 10 μl of antihuman CD73-APC and 10 μl of antihuman CD90-FITC were added into tubes labelled 1, 2, PerCP, PE, APC and FITC, respectively. Cells were mixed gently and incubated for 10 min in the dark at 4 °C.

Cells were washed by adding 1 ml of buffer and centrifuged at 300× g for 10 min. Supernatants were aspirated completely. Except for the tube labelled PerCP, cell pellet in all other tubes were resuspended into 500 μl of buffer and kept in a container filled with ice. 10 μl of anti-Biotin-PerCP was added to the tube labelled PerCP and cells were mixed gently. Following incubation for 10 min in the dark at 4 °C, cells were washed by adding 1 ml of buffer and centrifuged at 300× g for 10 min. Supernatant was aspirated completely and cells were resuspended in 500 μl buffer. Tubes labelled 1 and 2 were used to analyse the expression of specific antigen on SHED, and other five tubes were used to compensate the instrument. Multipotent (adipogenic, chondrogenic and osteogenic) differentiation potential was assessed by inducing the differentiation of SHED in StemPro® Adipogenesis Differentiation Kit (Gibco), StemPro® Osteogenesis Differentiation Kit (Gibco) and StemPro® Chondrogenesis Differentiation Kit (Gibco). For adipogenic and chondrogenic differentiation, cells were maintained in differentiation medium until day 14, and for osteogenic differentiation, cells were maintained until day 21. On every 3-day interval, cells were fed with fresh differentiation medium. Adipogenic, chondrogenic and osteogenic differentiation was confirmed by staining cells with Oil Red O (Sigma) solution, Safranin O (Sigma) solution and Alizarin Red S (Sigma) solution, respectively. For all differentiation assays, SHED cultured in complete proliferation medium were used as negative control.

2.4 Preparation of pHS

Venous blood was collected without any anticoagulant from healthy male donors aged 21–35 years ($n = 6$). Blood was collected from the donors having no record of smoking, alcohol consumption, drug and/narcotics addiction, inflammatory diseases either chronic or diagnosed within the past 4 weeks of blood collection, major surgical treatment in the last 1 year and immunotherapy. 20 ml of blood from each donor was coagulated separately in room temperature in sterile 50 ml centrifuge tube (BD Bioscience). The crude serum was separated from the coagulated blood by centrifugation initially at 400× g for 15 min and then at 1800× g for 15 min. The final serum was heat treated at 57 ± 2 °C for 30 min for complement inactivation. Individual heat-treated sera ($n = 6$) was pooled to prepare pHS for further experiments.

2.5 In Vitro Maintenance of SHED

Initial isolation and expansion of SHED ($n = 3$) were maintained in 10% FBS-supplemented KnockOut™ DMEM (Gibco) until passage

3. Subsequent cultures (from passage 4 to 7) were maintained in KnockOut™ DMEM (Gibco) supplemented with either 10% FBS or 10% pHS.

2.6 Effect of FBS and pHS on Proliferation of SHED

Live cell counts of SHED ($n = 3$) from passage 5 to passage 7 were taken in consideration to determine the effect of pHS-SM and FBS-SM on proliferation. Cells were counted by using trypan blue (Gibco) dye exclusion method. Three independent replications were performed to determine the population doubling (PD) of SHED from each donor ($n = 3$) at every passage. Data were analysed and plotted using Microsoft Excel. PD at each passage was calculated from the cell count by using the following equation:

$$X = \frac{\log_{10}(N_H) - \log_{10}(N_I)}{\log_{10}(2)}$$

where X = population doublings, N_I = inoculum number and N_H = cell harvest number. To obtain the cumulative population doubling (CPD), the PD increase at current passage was added to the PD of previous passages (Cristofalo et al. 1998; Li et al. 2015). Cumulative cell number was calculated by multiplying the initial seeding cell number (100,000 cells) at passage 4 by the CPD at subsequent passages (Dolley-Sonneville et al. 2013).

2.7 Effect of FBS and pHS on the Morphology of SHED

Photomicrographs of SHED ($n = 3$) were taken randomly at passage 5 on day 2 using inverted microscope (Primo Vert, Carl Zeiss) and analysed using ImageJ. In order to minimize the error and bias, all the cells present in one photomicrograph were counted. Clumped cells or cells without clear edge were excluded for this morphological study. After manually marking the surroundings of each cell's area, length and maximum Feret's diameter were measured using ImageJ (Haasters et al. 2009). Aspect ratio of each cell was determined by dividing Feret's diameter by the length of the cell.

2.8 Effect of FBS and pHS on Migration of SHED

Confluent SHED ($n = 3$) cultures at passage 5 were used for scratch assay. Pictures of the scratches were taken for 48 h at a 12-h interval under an inverted microscope (CKX41, Olympus, Centre Valley, USA). Average gap width of randomly selected three different points along the scratch was measured. Both gap width and the surface area of the gap were computed using Infinity Analyze (Lumenera Corporation, Ottawa, ON, Canada).

2.9 Paracrine Factor Analysis

Luminex-based ProcartaPlex human cytokine/chemokine 10plex immunoassay kit from eBioscience (Affymetrix, eBioscience, Vienna, Austria) was used to analyse the presence of the ten selected paracrine factors in each collected HS and FBS. These paracrine factors were selected on the basis of their involvement in regulating proliferation, migration and stemness of MSCs (Table S1).

2.10 Molecular Network Analysis

Ingenuity Pathway Analysis (IPA) (Ingenuity Systems; www.ingeniuty.com) was used to identify the role of the ten paracrine factors (using corresponding Entrez Gene IDs) in regulating biological functions and pathways.

2.11 Data Analysis

Data were analysed using SPSS version 22. The significant level was set at $p < 0.05$.

Table S1 List of selected ten paracrine factors that were analysed using ProcartaPlex human cytokine/chemokine 10plex immunoassay kit

Name of the paracrine factor [GenBank ID]	Function (References)
Epidermal growth factor (EGF) [NM_001178130]	Regulates proliferation of MSCs isolated from different origins while maintaining their regenerative potential (Tamama et al. 2006, 2010; Hu et al. 2013)
Fibroblast growth factor-2 (FGF-2) [NM_002006]	Stimulates the in vitro expansion of human BM-MSCs by activation of JNK signalling (Ahn et al. 2009)
	Slows down the ageing process of MSCs by decreasing the gradual loss of telomere sequences (Yanada et al. 2006; Bianchi et al. 2003)
	Cytoprotective role of FGFs have also been acknowledged by researchers (Werner and Grose 2003)
Granulocyte colony-stimulating factor (G-CSF) [NM_000759]	Promotes cellular proliferation and migration, and prevents apoptosis (Murakami et al. 2013)
	Mobilizes HSC and MSCs from bone marrow (Kawada et al. 2004)
	Improves chemotactic property of MSCs in vitro (Murakami et al. 2013)
Granulocyte macrophage colony-stimulating factor (GM-CSF) [NM_000758]	Acts as chemoattractant and induces mobilization of progenitors in the circulation (Rojas et al. 2005)
Hepatocyte growth factor (HGF) [NM_001010932]	Promotes proliferation and survival of various cell types (Forte et al. 2006)
	Induces migration and site-specific homing of various cell types including MSCs from different origins (Son et al. 2006)
Leukaemia inhibitory factor (LIF) [NM_002309]	Helps to maintain self-renewal and multidifferentiation potential of various stem cells including MSCs (Metcalf 2003; Kolf et al. 2007)
Platelet-derived growth factor BB (PDGF-BB) [NM_033016]	Induces both expansion and migration of MSCs (Fierro et al. 2007; Tamama et al. 2006)
	Helps survival of MSCs as well (Krausgrill et al. 2009)
Stem cell factor (SCF) [NM_000899]	Regulates the migration, differentiation and proliferation of several cell types (Lennartsson and Rönnstrand 2012)
	Induces the migration and homing of MSCs (Pan et al. 2013)
Stromal cell-derived factor-1α (SDF-1α) [NM_199168]	Helps site-specific migration and homing of MSCs through its receptor CXCR4 (He et al. 2010; Yu et al. 2015)
Vascular endothelial growth factor (VEGF) [NM_00117162]	Increases proliferation and survival of MSCs (Pons et al. 2008)

3 Results

3.1 SHED Share the Characteristics of MSCs

Homogenous monolayers of adherent and spindle-shaped SHED at different passages were detected under inverted microscope (Primo Vert, Carl Zeiss) (Fig. 1a–c). Flow cytometry data showed that 95% of the cultured SHED expressed the MSC positive markers (CD73, CD90, CD105) and 3% of the cultured SHED expressed the MSC negative markers (CD14, CD20, CD34, CD45) (Fig. 1d). Multipotent differentiation capability towards adipocytes, chondrocytes and osteocytes of the cultured SHED was observed with induced differentiation kit (Gibco) (Fig. 1e–g).

3.2 The pHS-SM Generates Adequate Number of SHED for Transplantation at Early Passage

Proliferation of SHED was significantly ($p < 0.05$) higher in pHS-SM compared to that in FBS-SM. Cumulative cell number showed that 1×10^6 viable SHED seeded at passage 4 could yield about 200×10^6 viable SHED at passage 7. Meanwhile, FBS-SM could yield 65×10^6

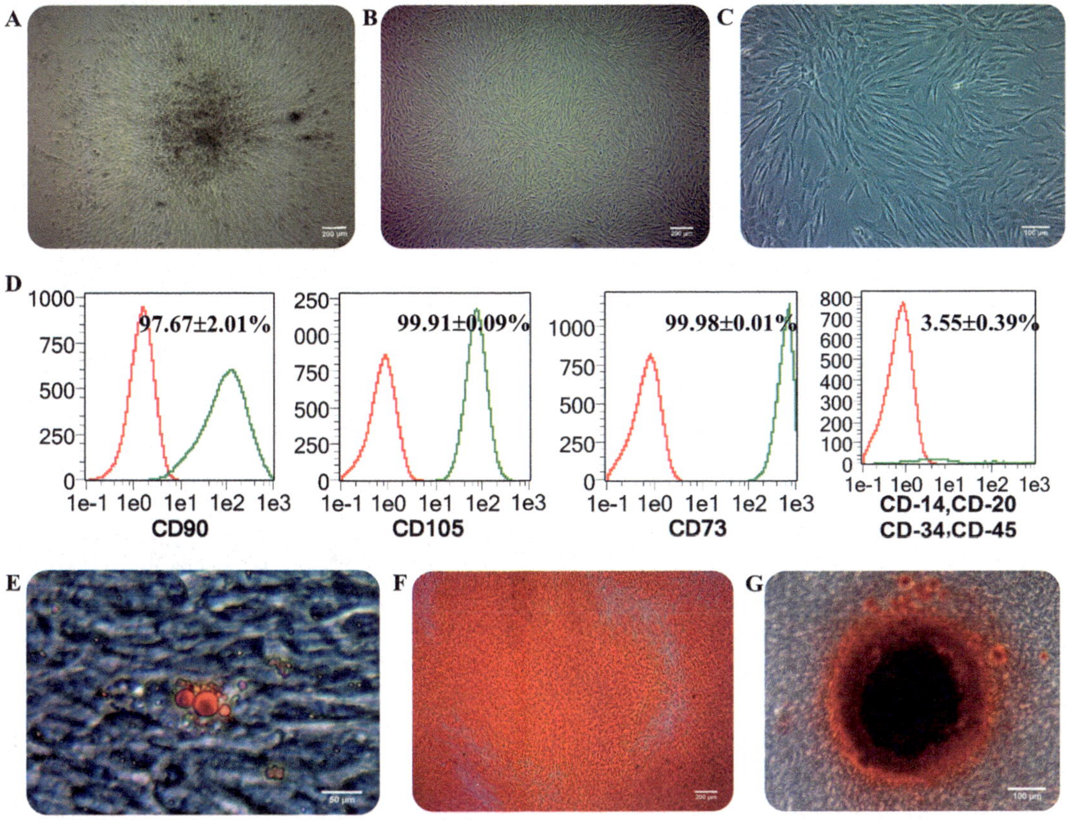

Fig. 1 Identification of SHED. (**a–c**) Plastic adherence of SHED. (**a**) SHED growing from primary dental pulp tissue on day 7. (**b**) Confluent homogenous monolayer of spindle-shaped SHED, passage 3. (**c**) Spindle-shaped SHED after 72 h of incubation, passage 5. (**d**) Expression of specific surface antigen on SHED. (**e–g**) Tri-lineage differentiation of SHED. SHED were cultured in chondrogenic and adipogenic differentiation medium for 14 days, and in osteogenic differentiation medium for 21 days. (**e**) After 14 days adipogenic differentiation, checked by staining with 'Oil red O'. (**f**) After 14 days chondrogenic differentiation, checked by staining the cells with 'Safranin-O'. (**g**) After 21 days osteogenic differentiation, checked by staining with 'Alizarin Red'. Photomicrographs (**a–c**, **e–g**) were taken using inverted microscope (Primo Vert, Carl Zeiss).

SHED only. The cumulative number of viable SHED from passage 5 to passage 7 in the pHS-SM found to be significantly higher compared to that in FBS-SM ($p < 0.01$) (Fig. 2a).

3.3 pHS Increase Homogeneity of SHED by Reducing Flat Cell

The morphological homogeneity was evaluated by measuring the number of cells having different morphology, cell surface area and aspect ratio. Number of FCs which appear to be partially differentiated and entered senescence was significantly lower ($p < 0.05$) in the pHS-SM (3%) compared to that in the FBS-SM (7%). However, no significant difference was observed between the proportion of RS cells in the FBS-SM and the pHS-SM (Fig. 2b). Surface area of FC subpopulations was in the range of 3018–12,499 μm^2 (Fig. 2c, d). Surface areas of other three subpopulations, i.e. RS, SS and FSS, were in the range of 202–2899, 566–2721 and 807–2939 μm^2, respectively. Notable, the average surface area (±SE) of the cell cultured in FBS-SM (1639 ± 63.66 μm^2) was significantly higher ($p = 0.023$) compared to that cultured in the pHS-SM (1354 ± 51.69 μm^2).

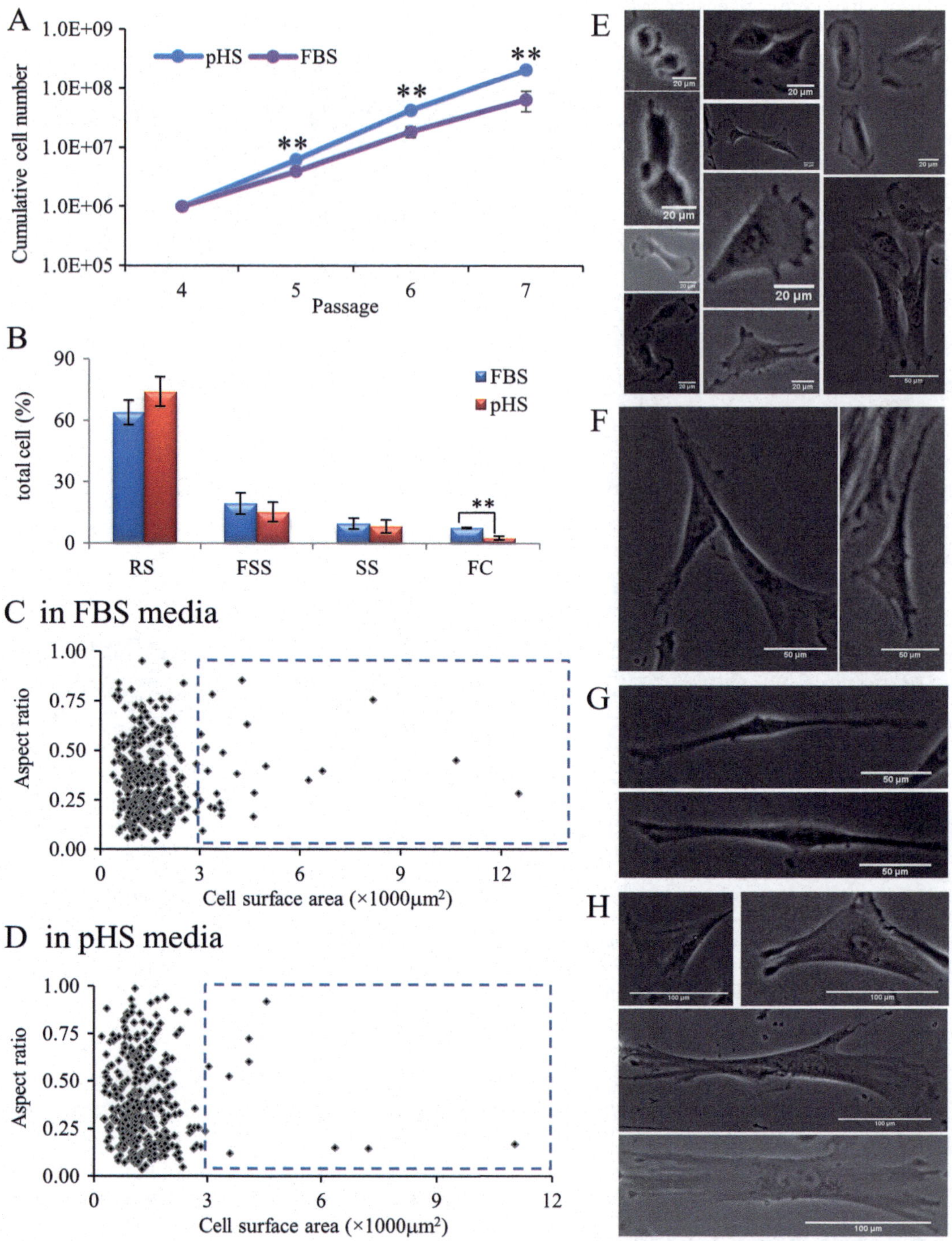

Fig. 2 Effect of FBS and pHS on the size and morphology of SHED. (**a**) Comparative growth of SHED cultured in FBS-SM and pHS-SM. (**b**) Percent distribution of SHED observed with different morphology in FBS and pHS-SM. (**c**) Dot plot graph shows the morphometric results of SHED in FBS-SM as area vs. aspect ratio. (**d**) Dot plot graph shows the morphometric results of SHED in pHS-SM as area vs. aspect ratio. (**e**) RS cell subpopulation of SHED with different morphology, (**f**) triangular spindle-shaped morphology of FSS cells, (**g**) elongated and spindle-shaped morphology of SS cells and (**h**) large and flattened morphology of FC cells ($n = 3$; ** significant at $p < 0.01$, * significant at $p < 0.05$; photomicrographs (**e–h**) were taken using inverted microscope, Primo Vert, Carl Zeiss). Scale bars: E, 20 μm; F and G, 50 μm; H, 100 μm

3.4 Migration of SHED in pHS Media Is More Directional

Migration of SHED during in vitro expansion was measured using the scratch assay. No significant difference in the closure of the gap width and surface area was observed when cultured either in FBS-SM or pHS-SM (Fig. 3a, b). However, migration of the SHED in pHS-SM appeared to be more directional (Fig. 3c).

3.5 Presence of Selected Paracrine Factors in HS

Presence of selected ten paracrine factors, namely, EGF, FGF-2, G-CSF, GM-CSF, HGF, LIF, PDGF-BB, SCF, SDF-1A and VEGF, was detected in the individual HS ($n = 6$) that were used to prepare pHS-SM (Table 1). However, none of these paracrine factors were detected in the FBS used to prepare FBS-SM.

3.6 Paracrine Factors Present in pHS Support Proliferation, Viability and Migration of SHED

IPA was performed to identify the possible signal transduction pathway that might have influenced the observed morphology, viability and migration of the SHED during the in vitro expansion in the pHS-SM. IPA predicted pathway analysis showed that the 'ephrin receptor signalling pathway' might have the influence on the higher proliferation and more directional migration of SHED cultured in pHS-SM (Fig. 4).

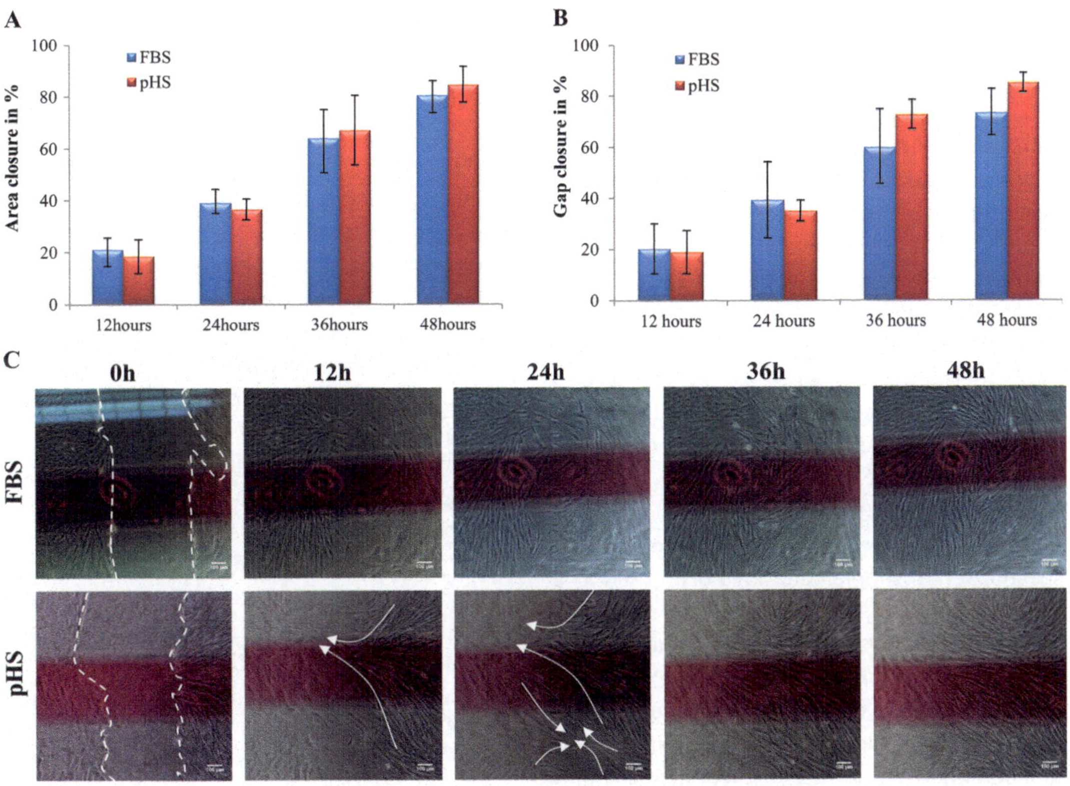

Fig. 3 Migration of SHED cultured in FBS and pHS-SM. (**a**) Closure of gap area by SHED that are cultured in FBS and pHS-SM. (**b**) Closure of gap width by SHED that are cultured in FBS and pHS-SM. (**c**) Representative photomicrographs of scratch assay ($n = 3$). *White arrows* indicate the directional migration of SHED. Scale bars: C, 100 μm

Table 1 Demographic profile and amount of selected paracrine factors in HS ($n = 6$) of each blood donor.

Variable	Donor 1	Donor 2	Donor 3	Donor 4	Donor 5	Donor 6	Range	Average
Blood group	A (+)ve	B (+)ve	B (+)ve	O (+)ve	O (+)ve	B (+)ve	NA	NA
Age (year)	32	29	25	28	28	30	25–32	28.7
Height (cm)	164	175	174	162	163	167	162–175	167.5
Weight (kg)	76	91	80	79	69	68	68–91	77.2
EGF (pg/ml)	55.7	20.9	14.0	40.6	81.1	32.0	14.0–81.1	40.7
FGF-2 (pg/ml)	0.0	0.0	0.0	0.0	56.8	0.0	0.0–56.8	9.5
G-CSF (pg/ml)	39.7	0.0	44.2	0.0	241.1	50.0	0.0–241.1	62.5
GM-CSF (pg/ml)	11.5	0.0	18.0	136.4	4.7	21.2	0.0–136	32.0
HGF (pg/ml)	42.2	135.2	64.6	44.2	254.4	19.4	19.4–254.4	93.4
LIF (pg/ml)	0.0	0.0	0.0	1.2	9.3	0.0	0.0–9.3	1.7
PDGF-BB (pg/ml)	507.1	265.1	167.2	284.5	127.8	107.0	107.0–507.1	243.1
SCF (pg/ml)	18.0	0.0	11.7	0.0	80.4	3.5	0.0–80.4	18.9
SDF-1A (pg/ml)	227.9	140.6	153.7	91.9	195.3	0.0	0.0–227.9	134.9
VEGF-A (pg/ml)	464.7	43.6	93.0	102.5	726.3	162.1	43.6–726.3	265.4

4 Discussion

It is practically critical to isolate a significantly higher number of stem cells, in the range of 200–400 million, from a single donor for transplantation (Pittenger et al. 1999). Hence, an in vitro expansion is inevitable prior to the transplantation of the isolated target stem cells, as the initially isolated number is far less than that is required (Pittenger et al. 1999; Haque et al. 2015). Routinely, in vitro culture media with FBS supplement is used for such in vitro expansion. However, a number of factors such as early cellular senescence and loss of migratory potential are common during the expansion of the stem cells in the FBS-SM (Estrada et al. 2012; Honczarenko et al. 2006; Bonab et al. 2006). Platelet lysate, plasma and ABO blood group-specific serum supplement have been used as an alternative to the FBS supplement during in vitro expansion (Kobayashi et al. 2005; Phadnis et al. 2006; Poloni et al. 2009; Tateishi et al. 2008; Blázquez-Prunera et al. 2017; Chase et al. 2010; Oikonomopoulos et al. 2015). Though proliferation potential varied with different media supplement, no significant differences in the differentiation potential and expression of MSC phenotypic markers were reported (Tateishi et al. 2008; Blázquez-Prunera et al. 2017; Chase et al. 2010; Oikonomopoulos et al. 2015). Hence, in this study, the potential of pHS in supporting the proliferation, homogeneity and migration of SHED was analysed. Successful isolation of MSCs from the bone marrow (Bieback et al. 2009) and adipose tissue (Paula et al. 2015) using AB blood group-specific pHS (AB-pHS) has been reported by several researchers. However, isolation of dental pulp-derived stem cells (DPSC) was not successful when the concentration of AB-pHS was below 20% (Khanna-Jain et al. 2012). Moreover, the volume of pulp within the deciduous teeth is not sufficient to culture in multiple media for isolation of SHED. Hence, to minimize donor to donor variations, all the SHED ($n = 3$) were first isolated and expanded (till passage 3) in FBS-supplemented media to yield enough cells needed for the subsequent in vitro experiments. It is noteworthy to mention that the xeno-contamination due to isolation and expansion in FBS media could be extremely reduced by subsequently culturing them in FBS-free medium only for 2 weeks (Komoda et al. 2010; Heiskanen et al. 2007).

Collection of blood irrespective of donors' blood group creates a larger donor pool compared to when the collection of blood is confined to a particular group. In search for a more

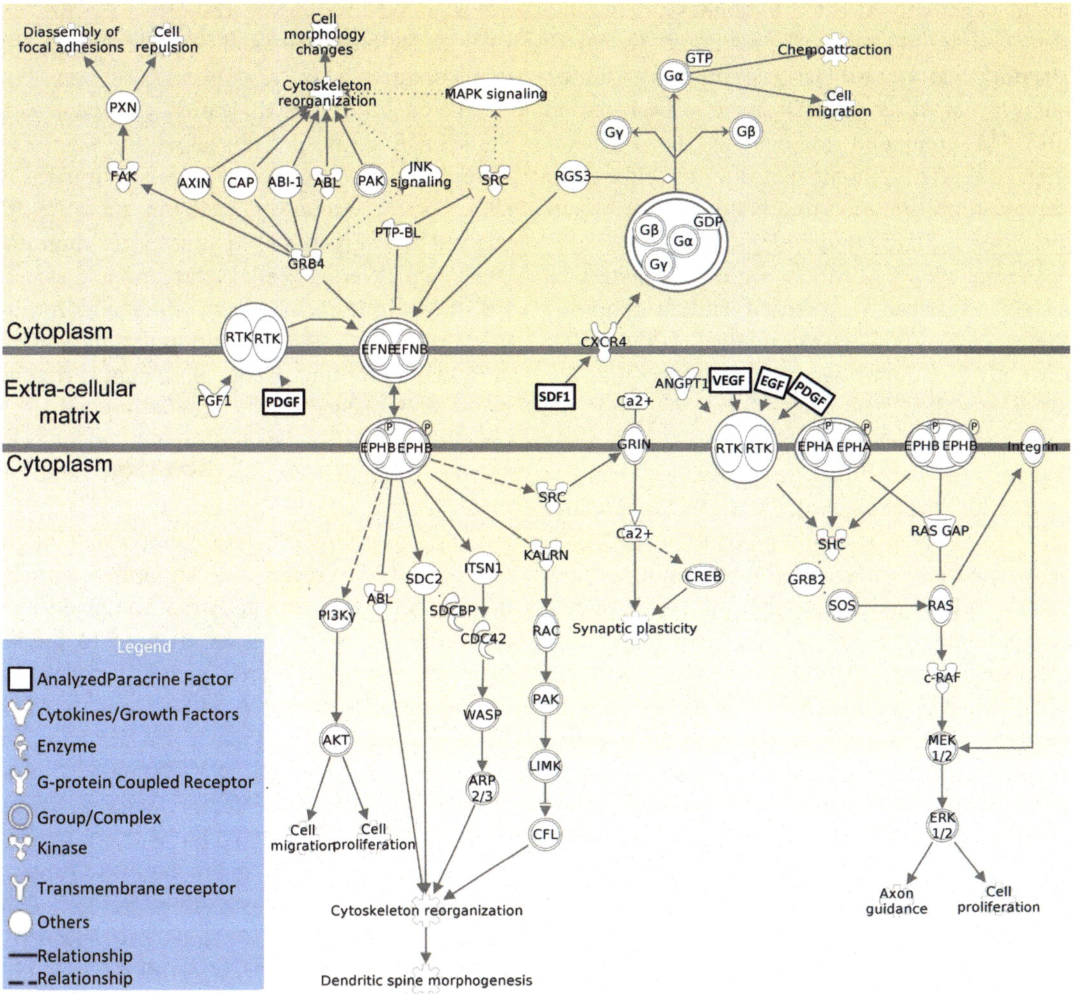

Fig. 4 Ephrin receptor signalling pathway. Paracrine factors analysed in the current study (EGF, PDGF-BB, SDF-1A and VEGF, shown in *black rectangle*) which were present in the pHS were predicted to regulate the proliferation, viability, migration and morphology of SHED through ephrin receptor signalling pathway (Ingenuity Systems; www.ingeniuty.com)

convenient alternative to FBS, in this study, we analysed the potential of pHS prepared using adult human serum of the donors, irrespective of the ABO blood grouping. However, the composition of individual HS differed markedly from individual to individual (Table 1). Therefore, sera from the six donors were pooled to prepare the pHS, as pooling the sera from a large number donors could help to minimize the batch to batch variation (Díez et al. 2015). Several techniques such as hypoxic culture conditions (1–2% O_2) and purified cytokines or growth factor supplement were attempted (Estrada et al. 2012; Saller et al. 2012; Bianchi et al. 2003; Strojny et al. 2015) to minimize ageing and loss of migratory properties of MSCs during in vitro expansion that are known to compromise their regenerative potential (Bonab et al. 2006). Hypoxic culture conditions provide better support for proliferation, yet the final outcome can be compromised due to the presence of xeno-antigen in FBS-SM (Haque et al. 2015). Use of a number of purified paracrine factors instead of FBS resulted in priming of MSCs in culture that might negatively affect the stemness, hence the regenerative potential (Prasanna et al. 2010; Handorf and Li 2011). A significantly higher number of live SHED were observed in pHS-SM compared to that in the FBS-SM (Fig. 2a). This might be due to either the increased proliferation or a better chance of survival of the SHED in the pHS-SM.

Again, ageing of MSCs is often characterized by its proliferation potential and morphology (Sethe et al. 2006; Haasters et al. 2009; Saller et al. 2012; Colter et al. 2001). Among the four different MSC subpopulations, FC and RS subpopulations showed the lowest and highest self-renewal capability, respectively (Docheva et al. 2008; Prockop et al. 2001). FC subpopulation from bone marrow (BM)-derived MSCs showed loss of differentiation potential as they were differentiated only to osteogenic progenitors or entered early phase of senescence (Matsuoka et al. 2013; Colter et al. 2001). In FBS-SM, BM-derived MSCs produced significantly higher number of RS cells in hypoxic condition (77%) compared to normoxic condition (67%) and lower number FC cells in hypoxia (1%) compared to that in normoxia (10%) (Saller et al. 2012).

The current study attests that the pHS-SM supports a higher morphological homogeneity of SHED with higher number of RS and lower number of FC in normoxic condition (Fig. 2b). During subculturing, the percentage of FC subpopulation was significantly lower, while the percentage of RS subpopulation was markedly higher in pHS-SM (Fig. 2b). Higher number of RS subpopulation and lower number of FC subpopulation in pHS-SM might be the clue behind the increased number of live SHED in this medium (Fig. 2a). Notably, for higher self-renewal capability and plasticity, cell population with higher number of RS cells is more preferable for regenerative therapy (Saller et al. 2012). Thus a more homogenous population of SHED in pHS-SM in adequate number, i.e. ~200 million at passage 7, denotes the suitability of using pHS-SM for in vitro expansion of MSCs before transplantation. However, further experiments are required to compare the effect of pHS and FBS on the morphological homogeneity in hypoxic condition.

The rate of in vitro migration, evaluated using the scratch assay, did not show any significant difference in relation to the FBS-SM or pHS-SM (Fig. 3a, b). However, cells in the pHS-SM showed a directional pattern of migration (Fig. 3c). Average cell surface area (±SE) in FBS-SM (1639 ± 63.66 μm^2) was significantly higher ($p = 0.026$) compared to that in pHS-SM (1354 ± 51.69 μm^2). The significant difference in the cell surface area might be related to the insignificant difference in scratch assay using SHED expanded in the FBS-SM and the pHS-SM.

Since proliferation, morphology and migration of SHED were found to be more favourable in pHS-SM, we compared the composition of paracrine factors in the serum used to prepare the pHS. Individual human serum that was used to prepare pHS as well as FBS was analysed for ten selected paracrine factors that are known to be involved in those biological actions (Table S1). All the selected paracrine factors were detected in individual human serum, while none of them were detected in the FBS (Table 1). Among the selected paracrine factors, EGF, FGF-2, G-CSF, GM-CSF, HGF, LIF, PDGF-BB and VEGF have proliferative effect on MSCs (Table S1), while FGF-2, G-CSF, GM-CSF, HGF, SCF, PDGF-BB and SDF-1A are known to enhance migration of MSCs (Table S1). Thus, the presence of paracrine factors in the pHS could be linked to the differences in biological functions such as proliferation, diversity in morphology and migration of SHED.

Biological functions and possible corresponding regulatory pathways were analysed by IPA in relation to the involvement of the selected paracrine factors. Among the IPA predicted pathways, 'ephrin receptor signalling pathway' was found to be involved in the regulation of proliferation, morphology and directional migration of SHED (Fig. 4). Notably the paracrine factors, namely, EGF, PDGF-BB, SDF-1A and VEGF, present in the pHS are known regulators in the ephrin receptor signalling pathway (Arvanitis and Davy 2008; Boyd et al. 2014). This predicted pathway also supports the hypothesis that the presence of these paracrine factors in pHS might be the cause behind the higher proliferation, homogeneity and migration of the SHED.

5 Conclusions

pHS has shown better effect on maintaining self-renewal capability and homogeneity of SHED. The effect of pHS on the other markers of ageing and migration could be studied for further confirmation. As it is not practical to use pHS throughout the process from isolation to transplantation of SHED or MSCs from other sources, we suggest that isolation can be done in the FBS-SM, and further expansion can be done in the pHS-SM. Furthermore, in vitro expansion in the pHS-SM before transplantation would minimize xeno-contamination from FBS, thus reducing chances of immune rejection. This novel finding though preliminary warrants further studies to confirm the effect of pHS in minimizing xeno-contamination during expansion of cells.

Acknowledgement This work was supported by High Impact Research MOHE Grant UM.C/625/1/HIR/MOHE/DENT/01 from the Ministry of Higher Education Malaysia. The authors thank the staff nurses from Oral and Maxillofacial Sciences Department, Faculty of Dentistry, University of Malaya, for their support in the collection of blood from donors.

Conflicts of Interest The authors deny any conflicts of interest related to this study.

References

Ahn, H.-J., Lee, W.-J., Kwack, K., & Kwon, Y. D. (2009). FGF2 stimulates the proliferation of human mesenchymal stem cells through the transient activation of JNK signaling. *FEBS Letters, 583*(17), 2922–2926.

Arvanitis, D., & Davy, A. (2008). Eph/ephrin signaling: Networks. *Genes & Development, 22*(4), 416–429.

Bianchi, G., Banfi, A., Mastrogiacomo, M., Notaro, R., Luzzatto, L., Cancedda, R., & Quarto, R. (2003). Ex vivo enrichment of mesenchymal cell progenitors by fibroblast growth factor 2. *Experimental Cell Research, 287*(1), 98–105.

Bieback, K., Hecker, A., Kocaömer, A., Lannert, H., Schallmoser, K., Strunk, D., & Klüter, H. (2009). Human alternatives to fetal bovine serum for the expansion of mesenchymal stromal cells from bone marrow. *Stem Cells, 27*.

Blázquez-Prunera, A., Díez, J. M., Gajardo, R., & Grancha, S. (2017). Human mesenchymal stem cells maintain their phenotype, multipotentiality, and genetic stability when cultured using a defined xeno-free human plasma fraction. *Stem Cell Research & Therapy, 8*, 103.

Bonab, M. M., Alimoghaddam, K., Talebian, F., Ghaffari, S. H., Ghavamzadeh, A., & Nikbin, B. (2006). Aging of mesenchymal stem cell in vitro. *BMC Cell Biology, 7*(1), 14.

Boyd, A. W., Bartlett, P. F., & Lackmann, M. (2014). Therapeutic targeting of EPH receptors and their ligands. *Nature Reviews. Drug Discovery, 13*(1), 39–62.

Chase, L. G., Lakshmipathy, U., Solchaga, L. A., Rao, M. S., & Vemuri, M. C. (2010). A novel serum-free medium for the expansion of human mesenchymal stem cells. *Stem Cell Research & Therapy, 1*(8), 1549–1553.

Colter, D. C., Sekiya, I., & Prockop, D. J. (2001). Identification of a subpopulation of rapidly self-renewing and multipotential adult stem cells in colonies of human marrow stromal cells. *Proceedings of the National Academy of Sciences of the United States of America, 98*(14), 7841–7845.

Cristofalo, V. J., Allen, R. G., Pignolo, R. J., Martin, B. G., & Beck, J. C. (1998). Relationship between donor age and the replicative lifespan of human cells in culture: A reevaluation. *Proceedings of the National Academy of Sciences of the United States of America, 95*(18), 10614–10619.

Díez, J. M., Bauman, E., Gajardo, R., & Jorquera, J. I. (2015). Culture of human mesenchymal stem cells using a candidate pharmaceutical grade xeno-free cell culture supplement derived from industrial human plasma pools. *Stem Cell Research & Therapy, 6*, 28.

Docheva, D., Padula, D., Popov, C., Mutschler, W., Clausen-Schaumann, H., & Schieker, M. (2008). Researching into the cellular shape, volume and elasticity of mesenchymal stem cells, osteoblasts and osteosarcoma cells by atomic force microscopy. *Journal of Cellular and Molecular Medicine, 12*(2), 537–552.

Dolley-Sonneville, P. J., Romeo, L. E., & Melkoumian, Z. K. (2013). Synthetic surface for expansion of human mesenchymal stem cells in xeno-free, chemically defined culture conditions. *PloS One, 8*(8), e70263.

Dominici, M., Le Blanc, K., Mueller, I., Slaper-Cortenbach, I., Marini, F., Krause, D., Deans, R., Keating, A., Prockop, D., & Horwitz, E. (2006). Minimal criteria for defining multipotent mesenchymal stromal cells. The International Society for Cellular Therapy position statement. *Cytotherapy, 8*(4), 315–317.

Eggenhofer, E., Luk, F., Dahlke, M. H., & Hoogduijn, M. J. (2014). The life and fate of mesenchymal stem cells. *Frontiers in Immunology, 5*, 148.

Estrada, J. C., Albo, C., Benguria, A., Dopazo, A., Lopez-Romero, P., Carrera-Quintanar, L., Roche, E., Clemente, E. P., Enriquez, J. A., Bernad, A., & Samper, E. (2012). Culture of human mesenchymal stem cells at low oxygen tension improves growth and genetic stability by activating glycolysis. *Cell Death and Differentiation, 19*(5), 743–755.

Eubanks, E. J., Tarle, S. A., & Kaigler, D. (2014). Tooth storage, dental pulp stem cell isolation, and clinical scale expansion without animal serum. *Journal of Endodontia, 40*(5), 652–657.

Fierro, F., Illmer, T., Jing, D., Schleyer, E., Ehninger, G., Boxberger, S., & Bornhäuser, M. (2007). Inhibition of platelet-derived growth factor receptorβ by imatinib mesylate suppresses proliferation and alters differentiation of human mesenchymal stem cells in vitro. *Cell Proliferation, 40*(3), 355–366.

Forte, G., Minieri, M., Cossa, P., Antenucci, D., Sala, M., Gnocchi, V., Fiaccavento, R., Carotenuto, F., De Vito, P., Baldini, P.M., Prat, M., & Di Nardo, P. (2006). Hepatocyte growth factor effects on mesenchymal stem cells: proliferation, migration, and differentiation. *Stem Cells, 24*(1), 23–33.

Govindasamy, V., Abdullah, A. N., Ronald, V. S., Musa, S., Ab Aziz, Z. A., Zain, R. B., Totey, S., Bhonde, R. R., & Abu Kasim, N. H. (2010). Inherent differential propensity of dental pulp stem cells derived from human deciduous and permanent teeth. *Journal of Endodontia, 36*(9), 1504–1515.

Haasters, F., Prall, W. C., Anz, D., Bourquin, C., Pautke, C., Endres, S., Mutschler, W., Docheva, D., & Schieker, M. (2009). Morphological and immunocytochemical characteristics indicate the yield of early progenitors and represent a quality control for human mesenchymal stem cell culturing. *Journal of Anatomy, 214*(5), 759–767.

Handorf, A. M., & Li, W. J. (2011). Fibroblast growth factor-2 primes human mesenchymal stem cells for enhanced chondrogenesis. *PloS One, 6*(7), e22887.

Haque, N., Kasim, N. H., & Rahman, M. T. (2015). Optimization of pre-transplantation conditions to enhance the efficacy of mesenchymal stem cells. *International Journal of Biological Sciences, 11*(3), 324–334.

He, X., Ma, J., & Jabbari, E. (2010). Migration of marrow stromal cells in response to sustained release of stromal-derived factor-1alpha from poly(lactide ethylene oxide fumarate) hydrogels. *International Journal of Pharmaceutics, 390*(2), 107–116.

Heiskanen, A., Satomaa, T., Tiitinen, S., Laitinen, A., Mannelin, S., Impola, U., Mikkola, M., Olsson, C., Miller-Podraza, H., Blomqvist, M., Olonen, A., Salo, H., Lehenkari, P., Tuuri, T., Otonkoski, T., Natunen, J., Saarinen, J., & Laine, J. (2007). N-glycolylneuraminic acid xenoantigen contamination of human embryonic and mesenchymal stem cells is substantially reversible. *Stem Cells, 25*(1), 197–202.

Honczarenko, M., Le, Y., Swierkowski, M., Ghiran, I., Glodek, A. M., & Silberstein, L. E. (2006). Human bone marrow stromal cells express a distinct set of biologically functional chemokine receptors. *Stem Cells, 24*(4), 1030–1041.

Hu, F., Wang, X., Liang, G., Lv, L., Zhu, Y., Sun, B., & Xiao, Z. (2013). Effects of epidermal growth gactor and basic fibroblast growth factor on the proliferation and osteogenic and neural differentiation of adipose-derived stem cells. *Cellular Reprogramming, 15*(3), 224–232.

Kawada, H., Fujita, J., Kinjo, K., Matsuzaki, Y., Tsuma, M., Miyatake, H., Muguruma, Y., Tsuboi, K., Itabashi, Y., Ikeda, Y., Ogawa, S., Okano, H., Hotta, T., Ando, K., & Fukuda, K. (2004). Nonhematopoietic mesenchymal stem cells can be mobilized and differentiate into cardiomyocytes after myocardial infarction. *Blood, 104*(12), 3581–3587.

Khanna-Jain, R., Vanhatupa, S., Vuorinen, A., Sandor, G., Suuronen, R., Mannerstrom, B., & Miettinen, S. (2012). Growth and differentiation of human dental pulp stem cells maintained in fetal bovine serum, human serum and serum-free/Xeno-free culture media. *Journal of Stem Cell Research & Therapy, 2*, 4.

Kobayashi, T., Watanabe, H., Yanagawa, T., Tsutsumi, S., Kayakabe, M., Shinozaki, T., Higuchi, H., & Takagishi, K. (2005). Motility and growth of human bone-marrow mesenchymal stem cells during ex vivo expansion in autologous serum. *Journal of Bone and Joint Surgery. British Volume (London), 87*(10), 1426–1433.

Kolf, C. M., Cho, E., & Tuan, R. S. (2007). Mesenchymal stromal cells. Biology of adult mesenchymal stem cells: regulation of niche, self-renewal and differentiation. *Arthritis Research & Therapy, 9*(1), 1–10.

Komoda, H., Okura, H., Lee, C. M., Sougawa, N., Iwayama, T., Hashikawa, T., Saga, A., Yamamoto-Kakuta, A., Ichinose, A., Murakami, S., Sawa, Y., & Matsuyama, A. (2010). Reduction of N-Glycolylneuraminic acid xenoantigen on human adipose tissue-derived stromal cells/mesenchymal stem cells leads to safer and more useful cell sources for various stem cell therapies. *Tissue Engineering – Part A, 16*(4), 1143–1155.

Krausgrill, B., Vantler, M., Burst, V., Raths, M., Halbach, M., Frank, K., Schynkowski, S., Schenk, K., Hescheler, J., Rosenkranz, S., & Müller-Ehmsen, J. (2009). Influence of cell treatment with PDGF-BB and reperfusion on cardiac persistence of mononuclear and mesenchymal bone marrow cells after transplantation into acute myocardial infarction in rats. *Cell Transplantation, 18*(8), 847–853.

Lee, R. H., Hsu, S. C., Munoz, J., Jung, J. S., Lee, N. R., Pochampally, R., & Prockop, D. J. (2006). A subset of human rapidly self-renewing marrow stromal cells preferentially engraft in mice. *Blood, 107*(5), 2153–2161.

Lennartsson, J., & Rönnstrand, L. (2012). Stem cell factor receptor/c-Kit: From basic science to clinical implications. *Physiological Reviews, 92*(4), 1619–1649.

Li, X., Hou, J., Wu, B., Chen, T., & Luo, A. (2014). Effects of platelet-rich plasma and cell coculture on angiogenesis in human dental pulp stem cells and endothelial progenitor cells. *Journal of Endodontia, 40*(11), 1810–1814.

Li, C.-Y., Wu, X.-Y., Tong, J.-B., Yang, X.-X., Zhao, J.-L., Zheng, Q.-F., Zhao, G.-B., & Ma, Z.-J. (2015). Comparative analysis of human mesenchymal stem cells from bone marrow and adipose tissue under xeno-free conditions for cell therapy. *Stem Cell Research & Therapy, 6*(1), 1–13.

Matsuoka, F., Takeuchi, I., Agata, H., Kagami, H., Shiono, H., Kiyota, Y., Honda, H., & Kato, R. (2013). Morphology-based prediction of osteogenic differentiation potential of human mesenchymal stem cells. *PloS One, 8*(2), e55082.

Metcalf, D. (2003). The unsolved enigmas of leukemia inhibitory factor. *Stem Cells, 21*(1), 5–14.

Murakami, M., Horibe, H., Iohara, K., Hayashi, Y., Osako, Y., Takei, Y., Nakata, K., Motoyama, N., Kurita, K., & Nakashima, M. (2013). The use of granulocyte-colony stimulating factor induced mobilization for isolation of dental pulp stem cells with high regenerative potential. *Biomaterials, 34*(36), 9036–9047.

Nakamura, S., Yamada, Y., Katagiri, W., Sugito, T., Ito, K., & Ueda, M. (2009). Stem cell proliferation pathways comparison between human exfoliated deciduous teeth and dental pulp stem cells by gene expression profile from promising dental pulp. *Journal of Endodontia, 35*(11), 1536–1542.

Oikonomopoulos, A., van Deen, W. K., Manansala, A. R., Lacey, P. N., Tomakili, T. A., Ziman, A., & Hommes, D. W. (2015). Optimization of human mesenchymal stem cell manufacturing: The effects of animal/xeno-free media. *Scientific Reports, 5*, 16570.

Pan, S., Dangaria, S., Gopinathan, G., Yan, X., Lu, X., Kolokythas, A., Niu, Y., & Luan, X. (2013). SCF promotes dental pulp progenitor migration, neovascularization, and collagen remodeling – potential applications as a homing factor in dental pulp regeneration. *Stem Cell Reviews & Reports, 9*(5), 655–667.

Paula, A. C. C., Martins, T. M. M., Zonari, A., Frade, S. P. P. J., Angelo, P. C., Gomes, D. A., & Goes, A. M. (2015). Human adipose tissue-derived stem cells cultured in xeno-free culture condition enhance c-MYC expression increasing proliferation but bypassing spontaneous cell transformation. *Stem Cell Research & Therapy, 6*(1), 76.

Phadnis, S. M., Joglekar, M. V., Venkateshan, V., Ghaskadbi, S. M., Hardikar, A. A., & Bhonde, R. R. (2006). Human umbilical cord blood serum promotes growth, proliferation, as well as differentiation of human bone marrow-derived progenitor cells. *In Vitro Cellular and Developmental Biology – Animal, 42*(10), 283–286.

Pittenger, M. F., Mackay, A. M., Beck, S. C., Jaiswal, R. K., Douglas, R., Mosca, J. D., Moorman, M. A., Simonetti, D. W., Craig, S., & Marshak, D. R. (1999). Multilineage potential of adult human mesenchymal stem cells. *Science, 284*(5411), 143–147.

Poloni, A., Maurizi, G., Rosini, V., Mondini, E., Mancini, S., Discepoli, G., Biasio, S., Battaglini, G., Felicetti, S., Berardinelli, E., Serrani, F., & Leoni, P. (2009). Selection of CD271(+) cells and human AB serum allows a large expansion of mesenchymal stromal cells from human bone marrow. *Cytotherapy, 11*(2), 153–162.

Pons, J., Huang, Y., Arakawa-Hoyt, J., Washko, D., Takagawa, J., Ye, J., Grossman, W., & Su, H. (2008). VEGF improves survival of mesenchymal stem cells in infarcted hearts. *Biochemical and Biophysical Research Communications, 376*(2), 419–422.

Prasanna, S. J., Gopalakrishnan, D., Shankar, S. R., & Vasandan, A. B. (2010). Pro-inflammatory cytokines, IFNgamma and TNFalpha, influence immune properties of human bone marrow and Wharton jelly mesenchymal stem cells differentially. *PloS One, 5*(2), e9016.

Prockop, D. J., Sekiya, I., & Colter, D. C. (2001). Isolation and characterization of rapidly self-renewing stem cells from cultures of human marrow stromal cells. *Cytotherapy, 3*(5), 393–396.

Rodrigues, M., Griffith, L., & Wells, A. (2010). Growth factor regulation of proliferation and survival of multipotential stromal cells. *Stem Cell Research & Therapy, 1*(4), 32.

Rojas, M., Xu, J., Woods, C. R., Mora, A. L., Spears, W., Roman, J., & Brigham, K. L. (2005). Bone marrow-derived mesenchymal stem cells in repair of the injured lung. *American Journal of Respiratory Cell and Molecular Biology, 33*(2), 145–152.

Saller, M. M., Prall, W. C., Docheva, D., Schonitzer, V., Popov, T., Anz, D., Clausen-Schaumann, H., Mutschler, W., Volkmer, E., Schieker, M., & Polzer, H. (2012). Increased stemness and migration of human mesenchymal stem cells in hypoxia is associated with altered integrin expression. *Biochemical and Biophysical Research Communications, 423*(2), 379–385.

Sethe, S., Scutt, A., & Stolzing, A. (2006). Aging of mesenchymal stem cells. *Ageing Research Reviews, 5*(1), 91–116.

Son, B. R., Marquez-Curtis, L. A., Kucia, M., Wysoczynski, M., Turner, A. R., Ratajczak, J., Ratajczak, M.Z., & Janowska-Wieczorek, A. (2006). Migration of bone marrow and cord blood mesenchymal stem cells in vitro is regulated by stromal-derived factor-1-CXCR4 and hepatocyte growth factor-c-met axes and involves matrix metalloproteinases. *Stem Cells, 24*(5), 1254–1264.

Strojny, C., Boyle, M., Bartholomew, A., Sundivakkam, P., & Alapati, S. (2015). Interferon gamma-treated dental pulp stem cells promote human mesenchymal stem cell migration in vitro. *Journal of Endodontia, 41*(8), 1259–1264.

Tamama, K., Fan, V. H., Griffith, L. G., Blair, H. C., & Wells, A. (2006). Epidermal growth factor as a candidate for ex vivo expansion of bone marrow–derived mesenchymal stem cells. *Stem Cells, 24*(3), 686–695.

Tamama, K., Kawasaki, H., & Wells, A. (2010). Epidermal growth factor (EGF) treatment on multipotential stromal cells (mscs). Possible enhancement of therapeutic potential of msc. *Journal of Biomedicine and Biotechnology, 2010*, 10.

Tateishi, K., Ando, W., Higuchi, C., Hart, D. A., Hashimoto, J., Nakata, K., Yoshikawa, H., & Nakamura, N. (2008). Comparison of human serum with fetal bovine serum for expansion and differentiation of human synovial MSC: Potential feasibility for clinical applications. *Cell Transplantation, 17*(5), 549–557.

Wang, X., Sha, X. J., Li, G. H., Yang, F. S., Ji, K., Wen, L. Y., Liu, S. Y., Chen, L., Ding, Y., & Xuan, K. (2012). Comparative characterization of stem cells from human exfoliated deciduous teeth and dental pulp stem cells. *Archives of Oral Biology, 57*(9), 1231–1240.

Werner, S., & Grose, R. (2003). Regulation of wound healing by growth factors and cytokines. *Physiological Reviews, 83*(3), 835–870.

Yanada, S., Ochi, M., Kojima, K., Sharman, P., Yasunaga, Y., & Hiyama, E. (2006). Possibility of selection of chondrogenic progenitor cells by telomere length in FGF-2-expanded mesenchymal stromal cells. *Cell Proliferation, 39*(6), 575–584.

Yu, Q., Liu, L., Lin, J., Wang, Y., Xuan, X., Guo, Y., & Hu, S. (2015). SDF-1α/CXCR4 axis mediates the migration of mesenchymal stem cells to the hypoxic-ischemic brain lesion in a rat model. *Cell Journal (Yakhteh), 16*(4), 440–447.

Advs Exp. Medicine, Biology - Innovations in Cancer Research and Regenerative Medicine (2018) 1: 45–62
DOI 10.1007/5584_2017_64

Published online: 21 July 2017

Role of Mesenchymal Stem Cells in Cancer Development and Their Use in Cancer Therapy

Nedime Serakinci, Pinar Tulay, and Rasime Kalkan

Abstract

Stem cells have the ability to perpetuate themselves through self-renewal and generate mature cells of a particular tissue through differentiation. Mesenchymal stem cells (MSCs) play an important role in tissue homeostasis – supporting tissue regeneration. MSCs are rare pluripotent cells supporting hematopoietic and mesenchymal cell lineages. MSCs have a great therapeutic potential in cancer therapy, as well as stem cell exosome and/or microvesicle-mediated tissue regeneration. In this review, the use of hMSCs in stem cell-mediated cancer therapy is discussed.

Keywords

Cancer development • Cancer therapy • Mesenchymal stem cells • Stem cell-mediated cancer therapy

Abbreviations

BM-MSC	Bone marrow-mesenchymal stem cells
GSCs	Germ line stem cells
HGF	Hepatocyte growth factor
IDO	Indoleamine 2,3-dioxygenase
IFNb	Interferon-b
MHC	Major histocompatibility complex
MSC	Mesenchymal stem cells
RA	Rheumatoid arthritis

1 Introduction

The development of multicellular organisms requires critical organization of events including

N. Serakinci (✉)
Faculty of Medicine, Department of Medical Genetics, Near East University, Nicosia, Cyprus

Faculty of Arts and Sciences, Department of Molecular Biology and Genetics, Near East University, Nicosia, Cyprus
e-mail: nedimeserakinci@gmail.com

P. Tulay and R. Kalkan
Faculty of Medicine, Department of Medical Genetics, Near East University, Nicosia, Cyprus
e-mail: pintulay@gmail.com; kalkanr@yahoo.com

cell division, cell proliferation, stem cell divisions and fate determination, and stem cell migration to specific niches and apoptosis. In adults, division and differentiation of a small number of stem cells in healthy tissues ensure a continuous turnover and optimal function of cells.

Stem cells have the ability to perpetuate themselves through self-renewal and generate mature cells of a particular tissue through differentiation. One of the distinct characteristics of stem cells is their asymmetric cell division in which one daughter cell remains to be undifferentiated, while the other has the ability to differentiate. The fate of stem cells is regulated with a balance between self-renewal and differentiation by both intrinsic and extrinsic signals from both the cell itself and the environment (Fig. 1) (Soltysova et al. 2005; Korbling et al. 2003). Stem cell pluripotency and self-renewal depend on multiple factors including the stem cell niche (Honoki et al. 2011; Morrison and Spradling 2008) and signaling pathways, such as Hedgehog, Notch, and Wnt/b-catenin (Soltysova et al. 2005). Any dysregulation or aberrant activation of these key pathways may result in the formation of cancer stem cells (CSCs) inducing tumorigenesis (Table 1) (Soltysova et al. 2005; Kitamura et al. 2009).

Stem cells have the capacity to divide for long periods of time in an environment where most of the cells are quiescent (Soltysova et al. 2005). The ability of adult human stem cells existence throughout the life of the organism is attributed to telomerase activity, and therefore, they actively maintain their telomere length to some degree (Kitamura et al. 2009). The ability to maintain telomere length allows them to have an extended proliferative capacity compared to other somatic cells (Kitamura et al. 2009). However, maintaining the telomeres of some adult stem cells may be limited, and therefore, it may not be sufficient to prevent their senescence. This may possess an increased risk of malignant transformation (Soltysova et al. 2005).

Stem cells can be classified into three categories: embryonic, germinal, and progenitor somatic stem cells. Embryonic stem (ES) cells are derived from the inner cell mass of the blastocyst, and they are the precursors of all cells of the organism with omnipotent stem cell characteristics. These cells are able to produce derivatives of all three germ layers (endoderm, mesoderm, and ectoderm) (Burdon et al. 2002).

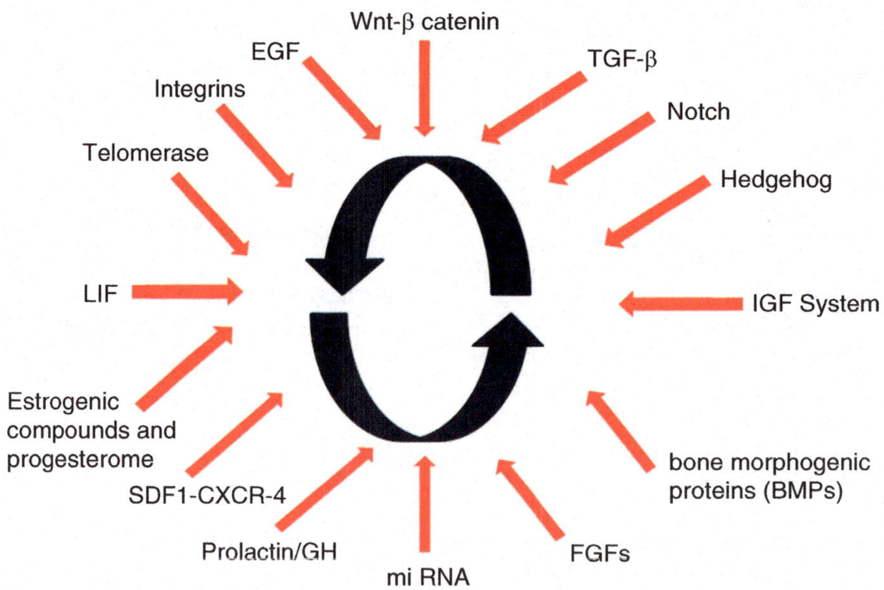

Fig. 1 Factors and signaling pathways involved in the regulation of self-renewal of stem cells
This figure shows the schematic diagram of the factors and signaling pathways that are shown to play a role in the regulation of stem cell self-renewal

ES cells have an indefinite replicative life span, which is attributable to expression of telomerase, like majority of cancer cells. ES cells differentiate to all cell lineages in vivo and can differentiate into many cell types in vitro (Soltysova et al. 2005; Hanna et al. 2010). Germ line stem cells (GSCs) can only produce gametes for reproduction. Since GSCs are responsible for passing on their genetic information from one generation to the next, sustaining their self-renewal ability is important to evolution and genetic continuity (Li and Xie 2005; Lin 2002). Somatic stem cells are multipotent and play essential roles in organogenesis and tissue maintenance. These cells are differentiated forms of ES cells. These cells are multipotent, and they are capable of self-renewal of a specific tissue or organ. One of the examples of a somatic stem cell is bone marrow. Adult bone marrow involves hematopoietic stem cells and other primitive progenitor cells often referred to as mesenchymal stem cells (MSCs).

MSCs play an important role in tissue homeostasis – supporting tissue regeneration. MSCs are rare pluripotent cells supporting hematopoietic and mesenchymal cell lineages including the bone, cartilage, fat, muscle, tendon, and marrow stroma (Soltysova et al. 2005; Van't Hof et al. 2007; Ji et al. 2009). These stem cells are considered to reside in a special microenvironment and are able to differentiate indefinitely. In recent years, histone modifications and other chromatin-based mechanisms are shown to be important for MSC differentiation capacity (Teven et al. 2011).

This chapter is focused on the double-faced characteristics of MSCs in cancer development and cancer therapy. The underlying mechanisms of double-faced characteristics of MSCs that is partly based on telomerase activity and maintenance of telomere homeostasis and thereby the cell's ability to be expanded in vitro are discussed; at the same time, potential use of MSCs in cancer therapeutics is evaluated in this chapter.

2 Mesenchymal Stem Cells

2.1 Characteristics of Mesenchymal Stem Cells

MSCs were first described by Friedenstein in 1968 (Friedenstein 1968). They were initially defined as fibroblast-like cells residing in the bone marrow. However, since there are a wide variety of sources for MSCs, different culture conditions including media and plating density introduce difficulties in establishing a solid-cut phenotype classification. Despite all these

Table 1 Comparison between somatic and cancer stem cell

Characteristics	Somatic stem cells	Cancer stem cells
Origin of cellular pedigree	+	+
Organ generation	+	
Tissue regeneration	+	
Cancer initiation		+
Self-renew	+	+
Pluripotency		
Tissue reconstruction	+	
Tumor formation		+
Long-lived	+	+
Resistant to damaging agents		
Anchorage-independent survival	+	+
Radioresistance chemoresistance		+

The table summarizes the common and different characteristics of somatic and cancer stem cells. The characteristic of each stem cell type is shown with '+'

complexities, recent reports define the phenotype of MSCs as large cells with prominent nucleoli and cells that are bleb-like and fibroblast-like, spindle-shaped, and flattened (Pevsner-Fischer et al. 2011).

MSCs have distinct characteristics and they can undergo self-renewing divisions, giving rise to progenitor cells. MSCs have the ability to differentiate into diverse tissue types of other lineages within or across germ lines including the mesodermal lineage, such as adipocytes, osteocytes, chondrocytes, and cells of other embryonic lineages (Jiang et al. 2002). MSCs secrete several paracrine factors including chemo-attractants for endothelial lineage cells, monocytes, and macrophages, and inflammatory factors, such as various chemokines and interleukins. Through the chemokine signaling, MSCs interact with the extracellular matrix that results in the transcription of target genes in cancer cells and macrophages and lymphocytes.

One of the main characteristics of hMSCs is their homing abilities to the primary tumor site and metastatic sites. Chemokines and their receptors were proposed to be involved in hMSC migration and homing (Kortesidis et al. 2005; Wynn et al. 2004). Moreover, recent studies have shown that hMSCs have antiapoptotic characteristics (Yang et al. 2014; Bhang et al. 2012) and bone marrow-derived MSCs (BM-MSCs) were shown to decrease oxidative stress, apoptosis and hippocampal damage in brain (Calio et al. 2014). Similar to the BM-MSCs, MSC-derived exosomes were shown to suppress inflammatory responses that repair tissue damage and modulate the immune system; however these findings remain controversial (Yu et al. 2014).

MSCs may interact with tumor cells to promote tumor growth directly or indirectly through autocrine/paracrine mechanisms. MSCs are considered to be the source of tumor-associated fibroblasts (TAFs) that are important components of tumor stroma. Therefore, MSCs play an important role in orchestrating the tumor microenvironment through angiogenesis and modulation of both immune system and tumor stromal architecture (Honoki et al. 2011). It is possible that MSCs provide a specific microenvironment or a niche for cancer stem cells. Therefore, investigation of interaction between stem cells and their specific microenvironment/niche cells will enhance the understanding of cancer development, especially metastasis (Sohni and Verfaillie 2013; Honoki et al. 2011; D'souza et al. 2012; Serakinci et al. 2004, 2011).

Several studies including animal models and preclinical investigations use MSCs in cancer treatments. BM-MSCs are mainly used for these studies; however these MSCs are not a very practical source since harvesting bone marrow is an invasive procedure that yields a small number of cells and differentiation potential and finally the lifespan of BM-MSCs reduces with the donors' age (Vellasamy et al. 2012; Bentzon et al. 2005; Kern et al. 2006; Mueller and Glowacki 2001; Serakinci et al. 2004, 2007). Although BM-MSCs are the gold standard for the in vitro experiments as well as clinical applications (Batsali et al. 2013), other alternative sources of MSCs, such as adipose tissue and umbilical cord blood, have gained more importance in the recent years. Adipose tissue-derived MSCs are obtained from the subcutaneous tissue and have similar expansion potential, differentiation capacity, and MSC immunophenotype as the MSCs derived from bone marrow (Kern et al. 2006). Umbilical cord, obtained after the removal of placenta, is rich in hematopoietic stem cells (Rubinstein et al. 1995; Wyrsch et al. 1999) and MSCs (Prindull et al. 1987). One of the differences between the MSCs obtained from bone marrow and umbilical cord is that umbilical cord-derived MSCs have a higher expansion rate compared to both bone marrow- and adipose tissue-derived MSCs (Kern et al. 2006; Goodwin et al. 2001), which could be due to the higher telomerase activity of the umbilical cord-derived MSCs (Chang et al. 2006).

2.2 Isolation of Mesenchymal Stem Cells

MSCs are present in several different types of tissues including adult and fetal tissues, the heart

(Chen et al. 2008), skeletal muscle (Peault et al. 2007), adipose tissue (Zuk et al. 2002), synovial tissue (De Bari et al. 2001), pancreas (Di Rocco et al. 2008), bone marrow (Friedenstein et al. 1970), umbilical cord blood (Erices et al. 2000), peripheral blood (Villaron et al. 2004; Fukuchi et al. 2004), and amniotic fluid (Tsai et al. 2004). MSCs have also been isolated from pathological tissues including rheumatoid arthritic joints (Marinova-Mutafchieva et al. 2000), and cells with similar characteristics as MSCs are present in most of the postnatal organs as well as tissues (Chamberlain et al. 2007). The isolation and enrichment of human MSCs (hMSCs) depend on culture media selection, their adhesion, and proliferation ability on tissue culture plastic in the presence of 10% fetal calf serum. The best way of isolation is immunopanning and combining several methods. One of the most widely used chemicals in the initial MSC isolation is the Ficoll™. Pre-enrichment strategies involve negative selection using different cell separation methodologies including the use of antibody cocktails to reduce the bone marrow-specific cell populations (Louis et al. 2001; Reyes et al. 2001) or via flow cytometry based on MSC surface proteins followed by confirmation of the cells under microscope (Campagnoli et al. 2001; Quirici et al. 2002; Li et al. 2006). During culturing, MSCs may experience a lag phase followed by rapid division. The in vitro doubling time of MSCs depend on the donor and the original plating density (Chamberlain et al. 2007).

MSCs derived from bone marrow have been attributed to have the highest degree of lineage plasticity, and following implantation into the early blastocysts, they have the capacity to give rise to all cell types (Jiang et al. 2002; Orlic et al. 2001). The aspirated bone marrow from the trabeculae of the bone is manipulated to remove the red blood cells, macrophages, and other extraneous materials (Boiret et al. 2005; Dennis et al. 2004; Pittenger et al. 1999; Terskikh et al. 2006; Serakinci et al. 2004). There is not an optimal antibody to define the MSCs. Therefore, MSC populations are isolated from human bone marrow using a range of composite cell surface phenotypes including STRO-1 monoclonal antibody in conjunction with antibodies against VCAM-1/CD106 (Simmons and Torok-Storb 1991), CD271 (Quirici et al. 2002), D7-Fib30, and CD49a.31 (Serakinci et al. 2004).

It is widely accepted that MSCs, especially BM-MSCs, are heterogeneous group of cells. This leads to difficulties in the identification of MSCs. The International Society for Cellular Therapy proposed the following minimal criteria for the correct characterization of hMSCs: adhesion of cells to plastic in standard culture conditions; differentiation of these cells into adipocytes, osteoblasts, and chondrocytes under standard in vitro differentiation conditions; and expression of prototypic hematopoietic antigens including SH2 (CD105), SH3/SH4 (CD73), CD29, CD44, CD90, CD71, CD106, CD166, STRO-1, GD2, and CD146 (Martinez et al. 2007; Pittenger et al. 1999; Shi and Gronthos 2003; Simmons and Torok-Storb 1991; Sordi et al. 2005) by more than 95% of the cells' population. On the other hand, cells must lack the expression of CD45, CD34, CD14 or CD11b, CD79a or CD19, and HLA class II markers (Dominici et al. 2006; Madrigal et al. 2014).

3 Proliferation Capacity of hMSC

After a limited number of cellular divisions, human somatic cells enter senescence, which is the phenomenon known as the Hayflick limit. Rb and p53 pathways play a significant role in senescence, and it is activated via the telomere signaling that has been strongly linked to this phenomenon. In the absence of Rb and p53 pathways, primary cells that continue cellular growth beyond Hayflick limit show severe telomere shortening, genetic instability, and cell death (Counter et al. 1992; Shay et al. 1991). With each cell division, telomeres shorten, and once they are critically shortened, the cell enters mortality stage 1 and goes into senescence (Kim et al. 1994). Cells that escape senescence and continue to proliferate enter either crises or mortality stage 2. If mutational changes occur and cells skip mortality stages 1 and 2, they may become immortal, which may lead to cancer. Telomerase reactivation in these cells plays a significant role in maintaining the telomeres at a constant length. Telomerase enzyme plays the most important role in the progressive shortening of telomeres and therefore has a vital role in cellular senescence.

hMSCs have stem cell characteristics; however, when they are forced to proliferate extensively, primary hMSC cultures quickly undergo replicative senescence with loss of proliferative potential with critically short telomeres (Bischoff et al. 2012). The replicative capacity of primary stem cells is limited, and the population doubling (PD) is around ten. This limited capacity restricts the use of stem cells in therapeutics. However, the limitation of proliferation capacity can be overcome by introduction of specific genes to enhance proliferation. One of these genes is the human telomerase reverse transcriptase (hTERT); in such hMSCs transduced with a lentivirus, expressing hTERT extends the proliferative capacity of stem cells and immortalizes the stem cells or progenitor cells (Bischoff et al. 2012; Serakinci et al. 2004, 2007; Simonsen et al. 2002). The telomerase-immortalized MSCs are cultured for 205 PDs, and the characteristic fibroblastic hMSC phenotype is maintained, suggesting a very low degree of random fluctuation in the telomere dynamics (Serakinci et al. 2007). Therefore, telomere dynamics have gained a great importance in stem cell function, especially in the expansion of stem cell populations.

Recent studies demonstrated that a prolonged culture of MSCs under low chronic stress increased the shortening rate of the mean telomere length, and they also showed that short-term, sublethal doses of the oxygen radical hydrogen peroxide caused significantly shortened telomeres (Harbo et al. 2012). When MSCs were exposed to hypoxic conditions, they exhibit greater colony-forming potential with faster and prolonged proliferation and differentiation and presented greater chemotaxis properties (Zeng et al. 2011). Hypoxia helps in maintaining the tumor stemlike cells in specific niches in which the self-renewal and differentiation activity is well balanced. Furthermore, recent studies showed that stemlike cells are localized in hypoxic zones of solid tumors in vivo (Das et al. 2008). Hypoxia also triggers the production of growth factors from MSC and allows the MSC to retain an undifferentiated phenotype, allowing for self-renewal without differentiation. This may be partially due to the fact that anatomically, MSCs tend to be found in hypoxic areas of the body (i.e., adipose tissue and bone marrow) and are relatively poorly perfused by the circulatory system (Madrigal et al. 2014).

4 Double-Faced Characteristics of MSCs: MSCs in Cancer Development and Therapeutic Applications

Several factors ranging from mitogens, extracellular matrix proteins, to angiogenic and inflammatory mediators can be produced directly by BM-MSCs or by other cells in the tumor microenvironment (Barcellos-De-Souza et al. 2013). This complex signaling cross talk may cause MSCs to have both stimulatory and inhibitory effects on tumor progression.

Recent advances showed that hMSC administration contributed to tumor development in animal models by promoting angiogenesis, creating a niche to support cancer stem cell survival, or having immunosuppressive ability (Momin et al. 2010; Sharma et al. 2009). On the contrary, it is widely acknowledged that hMSCs play an important role in cancer therapy due to their homing abilities to migrate to sites of injury, ischemia, and tumor microenvironments making them a good vehicle to be used in cancer therapy (Chamberlain et al. 2007; Brooke et al. 2007; Serakinci et al. 2011, 2014; Studeny et al. 2004). These controversies create double-faced characteristics of hMSCs in cancer.

4.1 MSCs in Cancer Development and Potential Neoplastic Transformation of hMSCs During Expansion

The interaction of MSCs with the stromal cells of the tumor might lead to MSC transformation. It is well acknowledged that cancer-associated stroma plays a role as much as the alterations of cancer cells in the tumor progression (Hanahan and Weinberg 2011). The tumor stroma consists of a compilation of cells and matrices, including fibroblasts/myofibroblasts, immune/inflammatory cells, blood vessels, connective tissues, and extracellular matrix. Tumor-associated stromal cells like fibroblasts/myofibroblasts, immune/inflammatory cells, and vascular endothelial cells are considered to be recruited from the surrounding normal tissue or from circulation and contribute to diverse aspects of tumor development and progression. Specialized MSCs in tumor stroma play an important role in the formation of tumor vessels.

The migration of MSCs toward the tumor microenvironment has been observed in many cancer types including lung (Loebinger et al. 2009), brain (Sasportas et al. 2009), breast (Patel et al. 2010), colon (Menon et al. 2007), pancreatic (Zischek et al. 2009), skin (Studeny et al. 2002), and ovarian (Kidd et al. 2009) cancers. Although the full mechanisms of how the MSCs migrate across endothelium and home to the target tissue are not well established, it is known that migration of MSCs depends on the different cytokine-receptor pairs, such as SDF-1/CXCR4, SCF-c-Kit, HGF/c-Met, VEGF/VEGFR, PDGF/PDGFr, MCP-1/CCR2, and HMGB1/RAGE (Momin et al. 2010). The potent pro-inflammatory cytokine, macrophage migration inhibitory factor (MIF), also plays a role in the migration of MSCs that is known to be produced by the tumor cells after the stimulation of MSCs (Barrilleaux et al. 2010). Additionally, MSCs stimulate tumor cells to produce a variety of cytokines and other growth factors (Yagi and Kitagawa 2013).

In the tumor microenvironment, BM-MSCs and their differentiated cells were shown to interact with the tumor cell as well as stromal cells through signaling molecules, such as autocrine/paracrine as well as angiogenesis, creating a complex cross talk (Sohni and Verfaillie 2013; Honoki et al. 2011; D'souza et al. 2012). It is possible that MSCs have a potential to provide a specific microenvironment or a niche for cancer stem cells. In addition, MSCs could be the origin of tumor-associated fibroblasts or myofibroblasts contributing to the formation of tumor microenvironment that eventually leads to the expansion and progression of tumors (Momin et al. 2010).

The pro-tumor behavior of BM-MSC may be caused by a combined effect of tumor proliferation, tumor dissemination, and vessel distribution (Spaeth et al. 2009; Coffelt et al. 2009; Zhu et al. 2006). The biological context including tumor histotypes or their local microenvironments may play an essential role in the balance between pro- and anti-survival effects for the tumor growth of BM-MSCs. By producing pro-angiogenic factors, such as VEGF-AA, IL-6, TGF-β, IL-8, leukemia inhibitory factor, macrophage-colony stimulating factor, and macrophage inflammatory protein-2, that activate angiogenesis, BM-MSCs may lead to tumor neo-angiogenesis by transdifferentiation into endothelial and/or pericytes-like cells (Suzuki et al. 2011; Spaeth et al. 2009).

The animal models play a major role in understanding the complex function of hMSC in cancer. The models generally rely on harvesting and culturing human tumor cells in vitro prior to grafting in animals. However these cells are not cultured with the tumor stroma and the epithelial cells. Therefore, even when these cells are injected into animals, the stroma is formed from the animals' own cells that may lead to the development of chimeric tumor. These interactions between MSCs and the tumor stroma may be investigated using various immunocompromised mice, such as NOG mice (Hahn and Weinberg 2002; Ito et al. 2002; Kim et al. 2004; Rangarajan and Weinberg 2003; Rosen and Jordan 2009). Mouse model of C57BL/6 infected by *Helicobacter felis* has shown that the development of gastric cancer was due to bone marrow stem cells tagged with beta galactosidase or GFP (Lazennec and Jorgensen 2008; Sarosi et al. 2008; Momin et al. 2010; Correa and Houghton 2007). Similar results of cancer development due to bone marrow stem cells were also reported in a rat model of Barrett's metaplasia (Sarosi et al. 2008). In glioma models, it was shown that BM-MSCs can integrate into tumor neovasculature, consequently forming as pericyte-like phenotype and acting as pericytes in tumor stroma (Bexell et al. 2009). The injection of BM-MSCs with cancer cells in immunocompromised animals led to increased tumor growth in models of B16 melanoma (Djouad et al. 2003), colon cancer (Shinagawa et al. 2010; Zhu et al. 2006), breast cancer (Karnoub et al. 2007; Liu et al. 2011), osteosarcoma (Bian et al. 2010), ovarian cancer (Spaeth et al. 2009), colorectal cancer (Tsai et al. 2011; De Boeck et al. 2013; Forbes et al. 2014), lung cancer (Suzuki et al. 2011), gastric cancer (Quante et al. 2011), and prostate carcinomas (Ye et al. 2012; Jung et al. 2013).

4.2 Applications of hMSCs in Cancer Therapy

To date, cancer therapy still remains to be one of the most challenging treatments. One of the reasons of this is that cancer has a dysregulated cellular self-renewal capacity. Gene and viral cancer therapies have shown improved outcomes; however, there is still a great need for development. Cancer therapy directed to tumor cells is very difficult, though the fundamental issue in cancer research is the identification of the cell type capable of sustaining the outgrowth of the neoplastic clone within solid tumors. Therefore therapies specifically directed to the cancer stem cells have gained importance to reduce and stop the metastatic tumors. Recent studies have shown that the use of stem cells obtained from adult tissue may be a novel vehicle for stem cell-mediated cancer therapy with improved antitumor effects.

Up to date, various agents have been used with stem cells as vehicles to reduce the tumor size or extend the survival of the organism. All these vehicles and agents showed different success rates (Momin et al. 2010; Aboody et al. 2008). In the last decade hMSCs have been proposed as a great tool in different therapeutic applications. MSCs serve as a powerful cell-based delivery vehicle for the site-specific release of therapeutic delivery of anticancer drugs due to their homing abilities, easy acquisition, hypoimmunogenic properties, fast ex vivo expansion, and feasibility of autologous transplantation properties (Gao et al. 2013). The use of hMSCs has an advantage over using other vehicles for delivery since the biological agents delivered within the patient's own bone marrow would not have a risk of being rejected as foreign objects by the immune system. Once these stem cells carrying biological agents are injected into the patient's bloodstream, they can migrate to the tumor site and release the anticarcinogenic agent.

The use of MSCs transduced with TRAIL showed induction of apoptosis and a subsequent reduction of tumor cell viability in colorectal carcinoma, gliomas, squamous cell carcinoma, and lung, breast, and cervical cancers (Shah 2013; Menon et al. 2009). Targeted delivery was also proven to be successful using hMSCs in xenogenic mouse model. The growth of malignant cells in the lungs of mice was inhibited both in vivo and in vitro following local delivery of MSCs transduced with IFN-β. However, inhibition has not been observed when the cells were administered systemically or directly to the tumor site (Studeny et al. 2002, 2004). In the prostate stroma of a castrate-resistant mouse model, MSCs were used to deliver frizzled related protein-2 (SFRP2) to antagonize the Wnt-mediated cancer progression by reducing tumor growth, increasing apoptosis, and potentially causing tumor necrosis (Placencio et al. 2010). In mice with both solid and metastatic tumors, it was shown that intratumoral injection of MSCs expressing modified interleukin-12 (MSCs/IL-12 M) caused strong tumor-specific T cell responses and antimetastatic effects as well as inhibitory effects of solid tumor growth. These effects were proven to be stronger than interleukin-12 expressing adenovirus (Seo et al. 2011). In 2011, Serakinci and colleagues showed the homing, engrafting, and proliferation abilities of hMSCs in a human xenograft model by transplanting an ovarian cancer cell line into immunocompromised mice (Serakinci et al. 2011). Human BM-MSCs were shown to secrete interferon-b (IFNb) and diminish melanoma, breast carcinoma, and lung metastases (Shah 2013). Similarly, MSCs derived from amniotic fluid were capable of transporting IFNb to the site of neoplasia of a bladder tumor model and inhibiting the tumor growth as well as prolonging the survival of mice (Shah 2013).

In addition to cancer treatments, MSCs have been used in many therapeutic applications including spinal cord injury (Kim et al. 2013a); bone disorders, such as osteogenesis imperfecta (OI) and hypophosphatasia (Kim and Cho 2013); cardiovascular diseases (Carvalho et al. 2013; Kim and Cho 2013; Hoogduijn et al. 2010); immune diseases (Cipriani et al. 2013); bone (Di Bella et al. 2008) and cartilage (Bulman et al. 2013) injury; and graft-versus-host disease due to bone marrow translation (Kim et al. 2013b). Additionally, intravenous allogeneic hMSCs home to the site of injury in human lung that was injured by *Escherichia coli* endotoxin (Serakinci et al. 2014). MSCs were successfully used to deliver prodrug-converting enzymes. One of the examples of this is the administration of ganciclovir via MSCs transduced with herpes simplex virus thymidine kinase (HSV-tk). This therapeutic regimen has been successfully employed in glioma and pancreatic cancer experimental models (Sun et al. 2011).

As mentioned earlier, one of other important factors of MSC relevant to therapeutic development relies on their anti-inflammatory and immunosuppressive properties. MSCs suppress T cell proliferation, B cell functions, natural killer cell proliferation, and cytokine production and prevent the differentiation, maturation, and activation of dendritic cells. MSCs can suppress cells independently of the major histocompatibility complex (MHC) identity between donor and recipient due to their low expression of MHC-II and other co-stimulatory molecules. MSCs can exert immunosuppressive effects by direct cell to cell contact; their primary mechanism is the production of soluble factors, such as transforming growth factor-β, hepatocyte growth factor (HGF), nitric oxide, and indoleamine 2,3-dioxygenase (IDO) (Kim and Cho 2013). Rheumatoid arthritis (RA) is a T cell-mediated autoimmune disease characterized by cartilage and bone destruction. Anti-inflammatory properties and regenerative potential of MSCs could offer a novel therapeutic approach to treat RA (Kim and Cho 2013).

4.3 Difficulties and Limiting Factors of MSC Use in Cancer Therapy

There are controversial issues of MSC use in cancer therapy. Difficulties starts already durring isolation process such as only 1 in every 10^5 cells obtained for the isolation constitutes of MSCs. Additionally, MSCs have low grafting efficiency as well as potency. The limited mitotic potential of hMSCs restrains their therapeutic applications especially since a high number of cells are required for therapy in humans. This raises the need for large-scale MSC expansion (Momin et al. 2010; Ma 2010). Telomere dynamics play an important role in stem cell function particularly in the expansion of stem cell population. Telomere homeostasis and telomerase play a critical role in tumor progression, and it is well known that the cancer cells rely on telomerase for its survival. One of the main functions of telomeres is to protect the chromosome ends as being detected as DNA double-strand breaks by DNA repair machinery.

Serakinci et al. (2007) have successfully introduced a retrovirus carrying hTERT gene and obtained an immortalized human MSCs (hMSC-telo1) cell line, which maintains the stem cell characteristics and has an expanded life span (Serakinci et al. 2007). This manipulation bypasses the naturally built-in controls of the cell that govern the delicate balance between cell proliferation, senescence, and carcinogenesis. Although this is very promising in the use of MSCs for therapeutic applications, manipulation of telomere-telomerase activity in order to extend the proliferative capacity of stem cell populations may increase the risk for stem cell susceptibility to carcinogenesis. The transduced cell line presented variations indicative of neoplastic development, such as contact inhibition, anchorage independence, and in vivo tumor formation in severe combined immunodeficiency (SCID) mice (Serakinci et al. 2004). These can be due to critically shortened telomeres leading to senescence that can be considered as a barrier against cancer formation via telomere-mediated checkpoint. Dysfunctional telomeres have the ability to disturb the genomic stability via Break-Fusion-Break cycles causing excessive genomic instability and aberrations, rapid cell proliferation, loss of contact inhibition, and gradual increase in telomerase activity (Serakinci et al. 2008; Furlani et al. 2009). These phenotypic and genotypic alterations were also reported in adipose-derived hMSCs (Rubio et al. 2005) and bone marrow-derived mouse MSCs (Miura et al. 2006; Tolar et al. 2007; Zhou et al. 2006). One of the options to resolve the issue of this neoplastic transformation of hMSCs after drug delivery could possibly be the self-destruction of the vehicle. This kind of treatment has been applied in tumor-selective viruses that mediate oncolytic effects on tumors and destroy targeted cancer cells. These kinds of viruses have been engineered based on telomerase promoter sequence with tumor-specific transcriptional response elements. These therapies target the telomerase-positive cells and combine the genetically engineered vehicle stem cell and suicide gene therapies (Abdul-Ghani et al. 2000; Bilsland et al. 2005; Komata et al. 2001; Plumb et al. 2001). Besides, Serakinci et al. (2008) have reported that hMSC-telo1 cells do not necessarily give rise to spontaneous transformation (Christensen et al. 2008). Neoplastic transformation was observed in the telomerase introduced hMSCs when they were subjected to 2.5Gy of gamma irradiation followed by long-term culturing. Thus the neoplastic transformation was suggested to occur due to DNA damage caused by irradiation and telomere damage leading to temporary cell cycle arrest (Serakinci et al. 2007; Christensen et al. 2008). Telomerase may help in the production of large number of cells, but it may have an impact on neoplastic transformation. Therefore these cells require close monitoring before and after the application and treatment.

In addition to the malignant transformations observed in cells with manipulated telomere-telomerase activity, unmanipulated BM-MSCs were also shown to produce a subpopulation of cells with high levels of telomerase activity, chromosomal aneuploidy, translocations, and capacity in the formation of tumors in multiple organs of NOD/SCID mice (Momin et al. 2010; Wang et al. 2005). Malignant transformations were also reported in rodent models, in mouse, and in hMSC populations (Rubio et al. 2008; Rubio et al. 2005; Zhou et al. 2006; Momin et al. 2010; Serakinci et al. 2004; Wang et al. 2005). The underlying molecular mechanism in this spontaneous transformation was suggested to occur after hMSC bypassed the senescence by repressing p16 levels followed by acquisition of telomerase activity, deletion at the Ink4a/Arf locus, and hyperphosphorylation of Rb (Rubio et al. 2008). This raised concerns that hMSCs can lead to spontaneous transformation when forced to extensive expansion. Contrary to these studies, the analysis of hMSCs with comparative genomic hybridization, karyotyping, and subtelomeric fluorescence in situ hybridization showed that there is no evidence of spontaneous hMSCs during long-term culture (Bernardo et al. 2007; Meza-Zepeda et al. 2008). However, maintaining a normal karyotype will not eliminate epigenetic changes, and telomerase-immortalized hMSCs were shown to accumulate genetic and epigenetic variations leading to spontaneous transformation (Serakinci et al. 2004; Burns et al. 2005). Exogenous administration of hMSCs may get engaged to developing tumors when infused systemically in animal models for glioma, colon carcinoma, gastric cancer, ovarian carcinoma, Kaposi's sarcoma, and melanoma (Correa and Houghton 2007; Lazennec and Jorgensen 2008). Although these studies supported the possibility of neoplastic transformation of hMSCs during in vitro expansion, it is still controversial, and the molecular pathogenesis underlying such mechanism is not fully established yet.

Moreover, a number of studies investigating the use of MSCs for graft-versus-host disease showed no signs of tumor formation (Correa and Houghton 2007; Barkholt et al. 2013; Resnick et al. 2013). A recent small-size phase 2 clinical study was reported for the use of MSCs for the treatment of Crohn's disease that showed promising results (Forbes et al. 2014). Although all these results are promising, in order to establish a definite role of MSC in tumorigenicity, comprehension of information on adult stem cells, regulation of their growth, and monitoring of the outcome of clinical applications are mandatory.

5 Conclusions and Future Directions

In the last decade, hMSCs have become a great therapeutic target for many diseases due to their homing abilities once reconstructed to inflammatory or tumor site. MSCs can be acquired from the patients' own body, and the use of these cells lowers the risks of rejection. In addition to their tumor-homing properties, MSCs are also easily transduced by integrating vectors due to their high levels of amphotropic receptors and offer long-term gene expression without alteration of phenotype. Gene- and viral-based therapies have shown enhancements in cancer treatment, and a number of anticancer genes have been successfully engineered into MSCs, which then promoted anticancer effects in various carcinoma models. However, since one of the most apparent characteristics of cancer is the continued cell growth that is associated with telomerase activity, there may be an increased risk of cancer development with the use of genetically modified cells in cancer therapies. Although one approach may be the use of telomerase inhibitors to activate cell death, these cells may still escape the cell death and cause genomic instability. Therefore, the use of targeted treatment with self-suicide vehicles may be a better approach to improve cancer therapy and reduce the risk of secondary tumors.

In all the studies using MSCs in delivery of therapeutic genes with lentivirus-mediated, retrovirus-mediated, plasmid-mediated and adenovirus to treat multiple diseases including cancer showed promising results in animal trials prolonging life of the animal, reducing complications, and/or tumor volume (Nguyen et al. 2014). The potential of MSCs for cell-based therapy has received tremendous attention. Transplantation of these cells has proven to be effective in treating a variety of genetic or acquired diseases due to their ability to engraft in various tissue types and differentiate into tissue-specific cells and release trophic factors to induce the tissue's own endogenous repair. However, the risk of MSC transformation that may lead to cancer development still remains ambiguous and needs to be further evaluated. Furthermore, poor engraftment and limited differentiation under in vivo conditions are major obstacles for efficient therapeutic use of MSCs. Therefore, although the use of MSCs holds a great promise in cancer therapies, the risk of not getting a treatment should be weighed against the risk of MSC transformation, and in vivo data supporting the true differentiation and regenerative potential of MSCs are still lacking. The success of these treatments depends on selecting an appropriate method for gene delivery to the cells, and the future treatment planning should be based on gene therapy and stem cell therapy combinations. Genetic modification of MSCs with beneficial genes of interest is a prerequisite for a successful use of stem cell-based therapeutic applications in the future.

In conclusion, the use of hMSCs is a promising approach for cancer therapy. The use of hMSCs in association with the conventional treatments may result in better prognosis of the patients. The use of hMSCs with suicidal gene therapies is a hopeful approach, and the combinatory effect of gene therapy with stem cell therapy may be the strategy forward; however the double-faced characteristics of hMSCs should always be kept in consideration.

References

Abdul-Ghani, R., Ohana, P., Matouk, I., Ayesh, S., Ayesh, B., Laster, M., Bibi, O., Giladi, H., Molnar-Kimber, K., Sughayer, M. A., De Groot, N., & Hochberg, A. (2000). Use of transcriptional regulatory sequences of telomerase (hTER and hTERT) for selective killing of cancer cells. *Molecular Therapy, 2*, 539–544.

Aboody, K. S., Najbauer, J., & Danks, M. K. (2008). Stem and progenitor cell-mediated tumor selective gene therapy. *Gene Therapy, 15*, 739–752.

Barcellos-De-Souza, P., Gori, V., Bambi, F., & Chiarugi, P. (2013). Tumor microenvironment: Bone marrow-mesenchymal stem cells as key players. *Biochimica et Biophysica Acta, 1836*, 321–335.

Barkholt, L., Flory, E., Jekerle, V., Lucas-Samuel, S., Ahnert, P., Bisset, L., Buscher, D., Fibbe, W., Foussat, A., Kwa, M., Lantz, O., Maciulaitis, R., Palomaki, T., Schneider, C. K., Sensebe, L., Tachdjian, G., Tarte, K., Tosca, L., & Salmikangas, P. (2013). Risk of tumorigenicity in mesenchymal stromal cell-based therapies–bridging scientific observations and regulatory viewpoints. *Cytotherapy, 15*, 753–759.

Barrilleaux, B. L., Fischer-Valuck, B. W., Gilliam, J. K., Phinney, D. G., & O'connor, K. C. (2010). Activation of CD74 inhibits migration of human mesenchymal stem cells. *In Vitro Cellular & Developmental Biology Animal*, 46, 566–572.

Batsali, A. K., Kastrinaki, M. C., Papadaki, H. A., & Pontikoglou, C. (2013). Mesenchymal stem cells derived from Wharton's Jelly of the umbilical cord: Biological properties and emerging clinical applications. *Current Stem Cell Research & Therapy, 8*, 144–155.

Bentzon, J. F., Stenderup, K., Hansen, F. D., Schroder, H. D., Abdallah, B. M., Jensen, T. G., & Kassem, M. (2005). Tissue distribution and engraftment of human mesenchymal stem cells immortalized by human telomerase reverse transcriptase gene. *Biochemical and Biophysical Research Communications, 330*, 633–640.

Bernardo, M. E., Emons, J. A., Karperien, M., Nauta, A. J., Willemze, R., Roelofs, H., Romeo, S., Marchini, A., Rappold, G. A., Vukicevic, S., Locatelli, F., & Fibbe, W. E. (2007). Human mesenchymal stem cells derived from bone marrow display a better chondrogenic differentiation compared with other sources. *Connective Tissue Research, 48*, 132–140.

Bexell, D., Gunnarsson, S., Tormin, A., Darabi, A., Gisselsson, D., Roybon, L., Scheding, S., & Bengzon, J. (2009). Bone marrow multipotent mesenchymal stroma cells act as pericyte-like migratory vehicles in experimental gliomas. *Molecular Therapy, 17*, 183–190.

Bhang, S. H., Lee, S., Shin, J. Y., Lee, T. J. & B.S., K. (2012) Transplantation of cord blood mesenchymal stem cells as spheroids enhances vascularization. Tissue Engineering. Part A, 18, 2138–2147.

Bian, Z. Y., Fan, Q. M., Li, G., Xu, W. T., & Tang, T. T. (2010). Human mesenchymal stem cells promote growth of osteosarcoma: Involvement of interleukin-6 in the interaction between human mesenchymal stem cells and Saos-2. *Cancer Science, 101*, 2554–2560.

Bilsland, A. E., Fletcher-Monaghan, A., & Keith, W. N. (2005). Properties of a telomerase specific Cre/Lox switch for transcriptionally targeted cancer gene therapy. *Neoplasia, 7*, 1020–1029.

Bischoff, D. S., Makhijani, N. S., & Yamaguchi, D. T. (2012). Constitutive expression of human telomerase enhances the proliferation potential of human mesenchymal stem cells. *BioResearch Open Access, 1*, 273–279.

Boiret, N., Rapatel, C., Veyrat-Masson, R., Guillouard, L., Guerin, J. J., Pigeon, P., Descamps, S., Boisgard, S., & Berger, M. G. (2005). Characterization of nonexpanded mesenchymal progenitor cells from normal adult human bone marrow. *Experimental Hematology, 33*, 219–225.

Brooke, G., Cook, M., Blair, C., Han, R., Heazlewood, C., Jones, B., Kambouris, M., Kollar, K., Mctaggart, S., Pelekanos, R., Rice, A., Rossetti, T., & Atkinson, K. (2007). Therapeutic applications of mesenchymal stromal cells. *Seminars in Cell & Developmental Biology, 18*, 846–858.

Bulman, S. E., Barron, V., Coleman, C. M., & Barry, F. (2013). Enhancing the mesenchymal stem cell therapeutic response: Cell localization and support for cartilage repair. *Tissue Engineering. Part B, Reviews, 19*, 58–68.

Burdon, T., Smith, A., & Savatier, P. (2002). Signalling, cell cycle and pluripotency in embryonic stem cells. *Trends in Cell Biology, 12*, 432–438.

Burns, J. S., Abdallah, B. M., Guldberg, P., Rygaard, J., Schroder, H. D., & Kassem, M. (2005). Tumorigenic heterogeneity in cancer stem cells evolved from long-term cultures of telomerase-immortalized human mesenchymal stem cells. *Cancer Research, 65*, 3126–3135.

Calio, M. L., Marinho, D. S., Ko, G. M., Ribeiro, R. R., Carbonel, A. F., Oyama, L. M., Ormanji, M., Guirao, T. P., Calio, P. L., Reis, L. A., Simoes Mde, J., Lisboa-Nascimento, T., Ferreira, A. T., & Bertoncini, C. R. (2014). Transplantation of bone marrow mesenchymal stem cells decreases oxidative stress, apoptosis, and hippocampal damage in brain of a spontaneous stroke model. *Free Radical Biology & Medicine, 70*, 141–154.

Campagnoli, C., Roberts, I. A., Kumar, S., Bennett, P. R., Bellantuono, I., & Fisk, N. M. (2001). Identification of mesenchymal stem/progenitor cells in human first-trimester fetal blood, liver, and bone marrow. *Blood, 98*, 2396–2402.

Carvalho, J. L., Braga, V. B., Melo, M. B., Campos, A. C., Oliveira, M. S., Gomes, D. A., Ferreira, A. J., Santos, R. A., & Goes, A. M. (2013). Priming mesenchymal stem cells boosts stem cell therapy to treat myocardial infarction. *Journal of Cellular and Molecular Medicine, 17*, 617–625.

Chamberlain, G., Fox, J., Ashton, B., & Middleton, J. (2007). Concise review: Mesenchymal stem cells: Their phenotype, differentiation capacity, immunological features, and potential for homing. *Stem Cells, 25*, 2739–2749.

Chang, Y. J., Shih, D. T., Tseng, C. P., Hsieh, T. B., Lee, D. C., & Hwang, S. M. (2006). Disparate mesenchyme-lineage tendencies in mesenchymal stem cells from human bone marrow and umbilical cord blood. *Stem Cells, 24*, 679–685.

Chen, Y., Shao, J. Z., Xiang, L. X., Dong, X. J., & Zhang, G. R. (2008). Mesenchymal stem cells: A promising candidate in regenerative medicine. *The International Journal of Biochemistry & Cell Biology, 40*, 815–820.

Christensen, R., Alsner, J., Brandt Sorensen, F., Dagnaes-Hansen, F., Kolvraa, S., & Serakinci, N. (2008) Transformation of human mesenchymal stem cells in radiation carcinogenesis: Long-term effect of ionizing radiation. *Regenerative Medicine, 3*, 849–861.

Cipriani, P., Di Benedetto, P., Liakouli, V., Del Papa, B., Di Padova, M., Di Ianni, M., Marrelli, A., Alesse, E., & Giacomelli, R. (2013). Mesenchymal stem cells (MSCs) from scleroderma patients (SSc) preserve their immunomodulatory properties although senescent and normally induce T regulatory cells (Tregs) with a functional phenotype: Implications for cellular-based therapy. *Clinical and Experimental Immunology, 173*, 195–206.

Coffelt, S. B., Marini, F. C., Watson, K., Zwezdaryk, K. J., Dembinski, J. L., Lamarca, H. L., Tomchuck, S. L., Honer Zu Bentrup, K., Danka, E. S., Henkle, S. L., & Scandurro, A. B. (2009). The pro-inflammatory peptide LL-37 promotes ovarian tumor progression through recruitment of multipotent mesenchymal stromal cells. *Proceedings of the National Academy of Sciences of the United States of America, 106*, 3806–3811.

Correa, P., & Houghton, J. (2007). Carcinogenesis of helicobacter pylori. *Gastroenterology, 133*, 659–672.

Counter, C. M., Avilion, A. A., Lefeuvre, C. E., Stewart, N. G., Greider, C. W., Harley, C. B., & Bacchetti, S. (1992). Telomere shortening associated with chromosome instability is arrested in immortal cells which express telomerase activity. *The EMBO Journal, 11*, 1921–1929.

Das, B., Tsuchida, R., Malkin, D., Koren, G., Baruchel, S., & Yeger, H. (2008). Hypoxia enhances tumor stemness by increasing the invasive and tumorigenic side population fraction. *Stem Cells, 26*, 1818–1830.

De Bari, C., Dell'accio, F., Tylzanowski, P., & Luyten, F. P. (2001). Multipotent mesenchymal stem cells from adult human synovial membrane. *Arthritis and Rheumatism, 44*, 1928–1942.

De Boeck, A., Pauwels, P., Hensen, K., Rummens, J. L., Westbroek, W., Hendrix, A., Maynard, D., Denys, H., Lambein, K., Braems, G., Gespach, C., Bracke, M., & De Wever, O. (2013). Bone marrow-derived mesenchymal stem cells promote colorectal cancer progression through paracrine neuregulin 1/HER3 signalling. *Gut, 62*, 550–560.

Dennis, J. E., Cohen, N., Goldberg, V. M., & Caplan, A. I. (2004). Targeted delivery of progenitor cells for cartilage repair. *Journal of Orthopaedic Research, 22*, 735–741.

Di Bella, C., Farlie, P., & Penington, A. J. (2008). Bone regeneration in a rabbit critical-sized skull defect using autologous adipose-derived cells. *Tissue Engineering. Part A, 14*, 483–490.

Di Rocco, G., Tritarelli, A., Toietta, G., Gatto, I., Iachininoto, M. G., Pagani, F., Mangoni, A., Straino, S., & Capogrossi, M. C. (2008). Spontaneous myogenic differentiation of Flk-1-positive cells from adult pancreas and other nonmuscle tissues. *American Journal of Physiology. Cell Physiology, 294*, C604–C612.

Djouad, F., Plence, P., Bony, C., Tropel, P., Apparailly, F., Sany, J., Noel, D., & Jorgensen, C. (2003). Immunosuppressive effect of mesenchymal stem cells favors tumor growth in allogeneic animals. *Blood, 102*, 3837–3844.

Dominici, M., Le Blanc, K., Mueller, I., Slaper-Cortenbach, I., Marini, F., Krause, D., Deans, R., Keating, A., Prockop, D., & Horwitz, E. (2006). Minimal criteria for defining multipotent mesenchymal stromal cells. The international society for cellular therapy position statement. *Cytotherapy, 8*, 315–317.

D'souza, N., Burns, J. S., Grisendi, G., Candini, O., Veronesi, E., Piccinno, S., Horwitz, E. M., Paolucci, P., Conte, P., & Dominici, M. (2012). MSC and tumors: Homing, differentiation, and secretion influence therapeutic potential. *Advances in Biochemical Engineering/Biotechnology, 130*, 209–266.

Erices, A., Conget, P., & Minguell, J. J. (2000). Mesenchymal progenitor cells in human umbilical cord blood. *British Journal of Haematology, 109*, 235–242.

Forbes, G. M., Sturm, M. J., Leong, R. W., Sparrow, M. P., Segarajasingam, D., Cummins, A. G., Phillips, M., & Herrmann, R. P. (2014). A phase 2 study of allogeneic mesenchymal stromal cells for luminal Crohn's disease refractory to biologic therapy. *Clinical Gastroenterology and Hepatology, 12*, 64–71.

Friedenstein, A. Y. (1968). Induction of bone tissue by transitional epithelium. *Clinical Orthopaedics and Related Research, 59*, 21–37.

Friedenstein, A. J., Chailakhjan, R. K., & Lalykina, K. S. (1970). The development of fibroblast colonies in monolayer cultures of guinea-pig bone marrow and spleen cells. *Cell and Tissue Kinetics, 3*, 393–403.

Fukuchi, Y., Nakajima, H., Sugiyama, D., Hirose, I., Kitamura, T., & Tsuji, K. (2004). Human placenta-derived cells have mesenchymal stem/progenitor cell potential. *Stem Cells, 22*, 649–658.

Furlani, D., Li, W., Pittermann, E., Klopsch, C., Wang, L., Knopp, A., Jungebluth, P., Thedinga, E., Havenstein, C., Westien, I., Ugurlucan, M., Li, R. K., Ma, N., & Steinhoff, G. (2009). A transformed cell population derived from cultured mesenchymal stem cells has no functional effect after transplantation into the injured heart. *Cell Transplantation, 18*, 319–331.

Gao, Z., Zhang, L., Hu, J., & Sun, Y. (2013). Mesenchymal stem cells: A potential targeted-delivery vehicle for anti-cancer drug, loaded nanoparticles. *Nanomedicine, 9*, 174–184.

Goodwin, H. S., Bicknese, A. R., Chien, S. N., Bogucki, B. D., Quinn, C. O., & Wall, D. A. (2001). Multilineage differentiation activity by cells isolated from umbilical cord blood: Expression of bone, fat, and neural markers. *Biology of Blood and Marrow Transplantation, 7*, 581–588.

Hahn, W. C., & Weinberg, R. A. (2002). Modelling the molecular circuitry of cancer. *Nature Reviews Cancer, 2*, 331–341.

Hanahan, D., & Weinberg, R. A. (2011). Hallmarks of cancer: The next generation. *Cell, 144*, 646–674.

Hanna, J., Cheng, A. W., Saha, K., Kim, J., Lengner, C. J., Soldner, F., Cassady, J. P., Muffat, J., Carey, B. W., & Jaenisch, R. (2010). Human embryonic stem cells with biological and epigenetic characteristics similar to those of mouse ESCs. *Proceedings of the National Academy of Sciences of the United States of America, 107*, 9222–9227.

Harbo, M., Koelvraa, S., Serakinci, N., & Bendix, L. (2012). Telomere dynamics in human mesenchymal stem cells after exposure to acute oxidative stress. *DNA Repair (Amst), 11*, 774–779.

Honoki, K., Fujii, H., & Tsujiuchi, T. (2011) *Cancer stem cell niche: The role of Mesenchymal stem cells in tumor microenvironment.*

Hoogduijn, M. J., Popp, F., Verbeek, R., Masoodi, M., Nicolaou, A., Baan, C., & Dahlke, M. H. (2010). The immunomodulatory properties of mesenchymal stem cells and their use for immunotherapy. *International Immunopharmacology, 10*, 1496–1500.

Ito, M., Hiramatsu, H., Kobayashi, K., Suzue, K., Kawahata, M., Hioki, K., Ueyama, Y., Koyanagi, Y., Sugamura, K., Tsuji, K., Heike, T., & Nakahata, T. (2002). NOD/SCID/gamma(c)(null) mouse: An excellent recipient mouse model for engraftment of human cells. *Blood, 100*, 3175–3182.

Ji, K. H., Xiong, J., Fan, L. X., Hu, K. M., & Liu, H. Q. (2009). Multilineage differentiation capability comparison between Mesenchymal stem cells and Multipotent adult progenitor cells. *Advanced Studies in Biology, 1*, 25–35.

Jiang, Y., Jahagirdar, B. N., Reinhardt, R. L., Schwartz, R. E., Keene, C. D., Ortiz-Gonzalez, X. R., Reyes, M., Lenvik, T., Lund, T., Blackstad, M., Du, J., Aldrich, S., Lisberg, A., Low, W. C., Largaespada, D. A., & Verfaillie, C. M. (2002). Pluripotency of mesenchymal stem cells derived from adult marrow. *Nature, 418*, 41–49.

Jung, Y., Kim, J. K., Shiozawa, Y., Wang, J., Mishra, A., Joseph, J., Berry, J. E., Mcgee, S., Lee, E., Sun, H., Jin, T., Zhang, H., Dai, J., Krebsbach, P. H., Keller, E. T., Pienta, K. J., & Taichman, R. S. (2013). Recruitment of mesenchymal stem cells into prostate tumours promotes metastasis. *Nature Communications, 4*, 1795.

Karnoub, A. E., Dash, A. B., Vo, A. P., Sullivan, A., Brooks, M. W., Bell, G. W., Richardson, A. L., Polyak, K., Tubo, R., & Weinberg, R. A. (2007). Mesenchymal stem cells within tumour stroma promote breast cancer metastasis. *Nature, 449*, 557–563.

Kern, S., Eichler, H., Stoeve, J., Kluter, H., & Bieback, K. (2006). Comparative analysis of mesenchymal stem cells from bone marrow, umbilical cord blood, or adipose tissue. *Stem Cells, 24*, 1294–1301.

Kidd, S., Spaeth, E., Dembinski, J. L., Dietrich, M., Watson, K., Klopp, A., Battula, V. L., Weil, M., Andreeff, M., & Marini, F. C. (2009). Direct evidence of mesenchymal stem cell tropism for tumor and wounding microenvironments using in vivo bioluminescent imaging. *Stem Cells, 27*, 2614–2623.

Kim, N., & Cho, S. G. (2013). Clinical applications of mesenchymal stem cells. *The Korean Journal of Internal Medicine, 28*, 387–402.

Kim, N. W., Piatyszek, M. A., Prowse, K. R., Harley, C. B., West, M. D., Ho, P. L., Coviello, G. M., Wright, W. E., Weinrich, S. L., & Shay, J. W. (1994). Specific association of human telomerase activity with immortal cells and cancer. *Science, 266*, 2011–2015.

Kim, J. B., O'hare, M. J., & Stein, R. (2004). Models of breast cancer: Is merging human and animal models the future? *Breast Cancer Research, 6*, 22–30.

Kim, J. W., Ha, K. Y., Molon, J. N., & Kim, Y. H. (2013a). Bone marrow-derived mesenchymal stem cell transplantation for chronic spinal cord injury in rats: Comparative study between intralesional and intravenous transplantation. *Spine (Phila Pa 1976), 38*, E1065–E1074.

Kim, N., Im, K. I., Lim, J. Y., Jeon, E. J., Nam, Y. S., Kim, E. J., & Cho, S. G. (2013b). Mesenchymal stem cells for the treatment and prevention of graft-versus-host disease: Experiments and practice. *Annals of Hematology, 92*, 1295–1308.

Kitamura, H., Okudela, K., Yazawa, T., Sato, H., & Shimoyamada, H. (2009). Cancer stem cell: Implications in cancer biology and therapy with special reference to lung cancer. *Lung Cancer, 66*, 275–281.

Komata, T., Kondo, Y., Kanzawa, T., Hirohata, S., Koga, S., Sumiyoshi, H., Srinivasula, S. M., Barna, B. P., Germano, I. M., Takakura, M., Inoue, M., Alnemri, E. S., Shay, J. W., Kyo, S., & Kondo, S. (2001). Treatment of malignant glioma cells with the transfer of constitutively active caspase-6 using the human telomerase catalytic subunit (human telomerase reverse transcriptase) gene promoter. *Cancer Research, 61*, 5796–5802.

Korbling, M., Estrov, Z., & Champlin, R. (2003). Adult stem cells and tissue repair. *Bone Marrow Transplantation, 32*(Suppl 1), S23–S24.

Kortesidis, A., Zannettino, A., Isenmann, S., Shi, S., Lapidot, T., & Gronthos, S. (2005). Stromal-derived factor-1 promotes the growth, survival, and development of human bone marrow stromal stem cells. *Blood, 105*, 3793–3801.

Lazennec, G., & Jorgensen, C. (2008). Concise review: Adult multipotent stromal cells and cancer: Risk or benefit? *Stem Cells, 26*, 1387–1394.

Li, L., & Xie, T. (2005). Stem cell niche: Structure and function. *Annual Review of Cell and Developmental Biology, 21*, 605–631.

Li, H., Fu, X., Ouyang, Y., Cai, C., Wang, J., & Sun, T. (2006). Adult bone-marrow-derived mesenchymal stem cells contribute to wound healing of skin appendages. *Cell and Tissue Research, 326*, 725–736.

Lin, H. (2002). The stem-cell niche theory: Lessons from flies. *Nature Reviews. Genetics, 3*, 931–940.

Liu, S., Ginestier, C., Ou, S. J., Clouthier, S. G., Patel, S. H., Monville, F., Korkaya, H., Heath, A., Dutcher, J., Kleer, C. G., Jung, Y., Dontu, G., Taichman, R., & Wicha, M. S. (2011). Breast cancer stem cells are regulated by mesenchymal stem cells through cytokine networks. *Cancer Research, 71*, 614–624.

Loebinger, M. R., Kyrtatos, P. G., Turmaine, M., Price, A. N., Pankhurst, Q., Lythgoe, M. F., & Janes, S. M. (2009). Magnetic resonance imaging of mesenchymal stem cells homing to pulmonary metastases using biocompatible magnetic nanoparticles. *Cancer Research, 69*, 8862–8867.

Louis, S. A., Zapf, R., Clarke, E., Thomas, T. E., & Sutherland, H. J. (2001). A negative-selection strategy for depleting myeloma cells from patients' BM and/or leukapheresis blood. *Cytotherapy, 3*, 489–504.

Ma, T. (2010). Mesenchymal stem cells: From bench to bedside. *World Journal of Stem Cells, 2*, 13–17.

Madrigal, M., Rao, K. S., & Riordan, N. H. (2014). A review of therapeutic effects of mesenchymal stem cell secretions and induction of secretory modification by different culture methods. *Journal of Translational Medicine, 12*, 260.

Marinova-Mutafchieva, L., Williams, R. O., Mauri, C., Mason, L. J., Walmsley, M. J., Taylor, P. C., Feldmann, M., & Maini, R. N. (2000). A comparative study into the mechanisms of action of anti-tumor necrosis factor alpha, anti-CD4, and combined anti-tumor necrosis factor alpha/anti-CD4 treatment in early collagen-induced arthritis. *Arthritis and Rheumatism, 43*, 638–644.

Martinez, C., Hofmann, T. J., Marino, R., Dominici, M., & Horwitz, E. M. (2007). Human bone marrow mesenchymal stromal cells express the neural ganglioside GD2: A novel surface marker for the identification of MSCs. *Blood, 109*, 4245–4248.

Menon, L. G., Picinich, S., Koneru, R., Gao, H., Lin, S. Y., Koneru, M., Mayer-Kuckuk, P., Glod, J., & Banerjee, D. (2007). Differential gene expression associated with migration of mesenchymal stem cells to conditioned medium from tumor cells or bone marrow cells. *Stem Cells, 25*, 520–528.

Menon, L. G., Kelly, K., Yang, H. W., Kim, S. K., Black, P. M., & Carroll, R. S. (2009). Human bone marrow-derived mesenchymal stromal cells expressing S-TRAIL as a cellular delivery vehicle for human glioma therapy. *Stem Cells, 27*, 2320–2330.

Meza-Zepeda, L. A., Noer, A., Dahl, J. A., Micci, F., Myklebost, O., & Collas, P. (2008). High-resolution analysis of genetic stability of human adipose tissue stem cells cultured to senescence. *Journal of Cellular and Molecular Medicine, 12*, 553–563.

Miura, M., Miura, Y., Padilla-Nash, H. M., Molinolo, A. A., Fu, B., Patel, V., Seo, B. M., Sonoyama, W., Zheng, J. J., Baker, C. C., Chen, W., Ried, T., & Shi, S. (2006). Accumulated chromosomal instability in murine bone marrow mesenchymal stem cells leads to malignant transformation. *Stem Cells (Dayton, Ohio), 24*(4), 1095–1103.

Momin, E. N., Vela, G., Zaidi, H. A., & Quinones-Hinojosa, A. (2010). The oncogenic potential of mesenchymal stem cells in the treatment of cancer: Directions for future research. *Current Immunology Reviews, 6*, 137–148.

Morrison, S. J., & Spradling, A. C. (2008). Stem cells and niches: Mechanisms that promote stem cell maintenance throughout life. *Cell, 132*(4), 598–611.

Mueller, S. M., & Glowacki, J. (2001). Age-related decline in the osteogenic potential of human bone marrow cells cultured in three-dimensional collagen sponges. *Journal of Cellular Biochemistry, 82*, 583–590.

Nguyen, S. T., Pham, V. Q., Phan, N. K., & Pham, P. V. (2014). Mesenchymal stem cell-based cancer gene therapy: Application and unresolved problems. *Annual Research & Review in Biology*, 4, 1387–1396.

Orlic, D., Kajstura, J., Chimenti, S., Jakoniuk, I., Anderson, S. M., Li, B., Pickel, J., Mckay, R., Nadal-Ginard, B., Bodine, D. M., Leri, A., & Anversa, P. (2001). Bone marrow cells regenerate infarcted myocardium. Nature, 410, 701–705.

Patel, S. A., Meyer, J. R., Greco, S. J., Corcoran, K. E., Bryan, M., & Rameshwar, P. (2010). Mesenchymal stem cells protect breast cancer cells through regulatory T cells: Role of mesenchymal stem cell-derived TGF-beta. *Journal of Immunology, 184*, 5885–5894.

Peault, B., Rudnicki, M., Torrente, Y., Cossu, G., Tremblay, J. P., Partridge, T., Gussoni, E., Kunkel, L. M., & Huard, J. (2007). Stem and progenitor cells in skeletal muscle development, maintenance, and therapy. *Molecular Therapy, 15*, 867–877.

Pevsner-Fischer, M., Levin, S., & Zipori, D. (2011). The origins of mesenchymal stromal cell heterogeneity. *Stem Cell Reviews, 7*, 560–568.

Pittenger, M. F., Mackay, A. M., Beck, S. C., Jaiswal, R. K., Douglas, R., Mosca, J. D., Moorman, M. A., Simonetti, D. W., Craig, S., & Marshak, D. R. (1999).

Multilineage potential of adult human mesenchymal stem cells. *Science, 284*, 143–147.

Placencio, V. R., Li, X., Sherrill, T. P., Fritz, G., & Bhowmick, N. A. (2010). Bone marrow derived mesenchymal stem cells incorporate into the prostate during regrowth. *PloS One, 5*, e12920.

Plumb, J. A., Bilsland, A., Kakani, R., Zhao, J., Glasspool, R. M., Knox, R. J., Evans, T. R., & Keith, W. N. (2001). Telomerase-specific suicide gene therapy vectors expressing bacterial nitroreductase sensitize human cancer cells to the pro-drug CB1954. *Oncogene, 20*, 7797–7803.

Prindull, G., Ben-Ishay, Z., Ebell, W., Bergholz, M., Dirk, T., & Prindull, B. (1987). CFU-F circulating in cord blood. *Blut, 54*, 351–359.

Quante, M., Tu, S. P., Tomita, H., Gonda, T., Wang, S. S., Takashi, S., Baik, G. H., Shibata, W., Diprete, B., Betz, K. S., Friedman, R., Varro, A., Tycko, B., & Wang, T. C. (2011). Bone marrow-derived myofibroblasts contribute to the mesenchymal stem cell niche and promote tumor growth. *Cancer Cell, 19*, 257–272.

Quirici, N., Soligo, D., Bossolasco, P., Servida, F., Lumini, C., & Deliliers, G. L. (2002). Isolation of bone marrow mesenchymal stem cells by anti-nerve growth factor receptor antibodies. *Experimental Hematology, 30*, 783–791.

Rangarajan, A., & Weinberg, R. A. (2003). Opinion: Comparative biology of mouse versus human cells: Modelling human cancer in mice. *Nature Reviews Cancer, 3*, 952–959.

Resnick, I. B., Barkats, C., Shapira, M. Y., Stepensky, P., Bloom, A. I., Shimoni, A., Mankuta, D., Varda-Bloom, N., Rheingold, L., Yeshurun, M., Bielorai, B., Toren, A., Zuckerman, T., Nagler, A. & Or, R. (2013) Treatment of severe steroid resistant acute GVHD with mesenchymal stromal cells (MSC). American Journal of Blood Research, 3**,** 225–238.

Reyes, M., Lund, T., Lenvik, T., Aguiar, D., Koodie, L., & Verfaillie, C. M. (2001). Purification and ex vivo expansion of postnatal human marrow mesodermal progenitor cells. *Blood, 98*, 2615–2625.

Rosen, J. M., & Jordan, C. T. (2009). The increasing complexity of the cancer stem cell paradigm. *Science, 324*, 1670–1673.

Rubinstein, P., Dobrila, L., Rosenfield, R. E., Adamson, J. W., Migliaccio, G., Migliaccio, A. R., Taylor, P. E., & Stevens, C. E. (1995). Processing and cryopreservation of placental/umbilical cord blood for unrelated bone marrow reconstitution. *Proceedings of the National Academy of Sciences of the United States of America, 92*, 10119–10122.

Rubio, D., Garcia-Castro, J., Martin, M. C., De La Fuente, R., Cigudosa, J. C., Lloyd, A. C., & Bernad, A. (2005). Spontaneous human adult stem cell transformation. *Cancer Research, 65*, 3035–3039.

Rubio, D., Garcia, S., Paz, M. F., De La Cueva, T., Lopez-Fernandez, L. A., Lloyd, A. C., Garcia-Castro, J., & Bernad, A. (2008). Molecular characterization of spontaneous 17 mesenchymal stem cell transformation. *PloS One, 3*, e1398.

Sarosi, G., Brown, G., Jaiswal, K., Feagins, L. A., Lee, E., Crook, T. W., Souza, R. F., Zou, Y. S., Shay, J. W., & Spechler, S. J. (2008). Bone marrow progenitor cells contribute to esophageal regeneration and metaplasia in a rat model of Barrett's esophagus. *Diseases of the Esophagus, 21*, 43–50.

Sasportas, L. S., Kasmieh, R., Wakimoto, H., Hingtgen, S., Van De Water, J. A., Mohapatra, G., Figueiredo, J. L., Martuza, R. L., Weissleder, R., & Shah, K. (2009). Assessment of therapeutic efficacy and fate of engineered human mesenchymal stem cells for cancer therapy. *Proceedings of the National Academy of Sciences of the United States of America, 106*, 4822–4827.

Seo, S. H., Kim, K. S., Park, S. H., Suh, Y. S., Kim, S. J., Jeun, S. S., & Sung, Y. C. (2011). The effects of mesenchymal stem cells injected via different routes on modified IL-12-mediated antitumor activity. *Gene Therapy, 18*, 488–495.

Serakinci, N., Guldberg, P., Burns, J. S., Abdallah, B., Schrodder, H., Jensen, T., & Kassem, M. (2004). Adult human mesenchymal stem cell as a target for neoplastic transformation. *Oncogene, 23*, 5095–5098.

Serakinci, N., Christensen, R., Graakjaer, J., Cairney, C. J., Keith, W. N., Alsner, J., Saretzki, G., & Kolvraa, S. (2007). Ectopically hTERT expressing adult human mesenchymal stem cells are less radiosensitive than their telomerase negative counterpart. *Experimental Cell Research, 313*, 1056–1067.

Serakinci, N., Graakjaer, J., & Kolvraa, S. (2008). Telomere stability and telomerase in mesenchymal stem cells. *Biochimie, 90*, 33–40.

Serakinci, N., Christensen, R., Fahrioglu, U., Sorensen, F. B., Dagnæs-Hansen, F., Hajek, M., Jensen, T. H., Kolvraa, S., & Keith, N. W. (2011). Mesenchymal stem cells as therapeutic delivery vehicles targeting tumor stroma. *Cancer Biotherapy & Radiopharmaceuticals, 26*, 767–773.

Serakinci, N., Fahrioglu, U., & Christensen, R. (2014). Mesenchymal stem cells, cancer challenges and new directions. *European Journal of Cancer, 50*, 1522–1530.

Shah, K. (2013). Encapsulated stem cells for cancer therapy. *Biomatter, 3*(1), e24278.

Sharma, A. K., Fuller, N. J., Sullivan, R. R., Fulton, N., Hota, P. V., Harrington, D. A., Villano, J., Hagerty, J. A., & Cheng, E. Y. (2009). Defined populations of bone marrow derived mesenchymal stem and endothelial progenitor cells for bladder regeneration. *The Journal of Urology, 182*, 1898–1905.

Shay, J. W., Pereira-Smith, O. M., & Wright, W. E. (1991). A role for both RB and p53 in the regulation of human cellular senescence. *Experimental Cell Research, 196*, 33–39.

Shi, S., & Gronthos, S. (2003). Perivascular niche of postnatal mesenchymal stem cells in human bone marrow and dental pulp. *Journal of Bone and Mineral Research, 18*, 696–704.

Shinagawa, K., Kitadai, Y., Tanaka, M., Sumida, T., Kodama, M., Higashi, Y., Tanaka, S., Yasui, W., & Chayama, K. (2010). Mesenchymal stem cells enhance growth and metastasis of colon cancer. *International Journal of Cancer, 127*, 2323–2333.

Simmons, P. J., & Torok-Storb, B. (1991). Identification of stromal cell precursors in human bone marrow by a novel monoclonal antibody, STRO-1. *Blood, 78*, 55–62.

Simonsen, J. L., Rosada, C., Serakinci, N., Justesen, J., Stenderup, K., Rattan, S. I., Jensen, T. G., & Kassem, M. (2002). Telomerase expression extends the proliferative life-span and maintains the osteogenic potential of human bone marrow stromal cells. *Nature Biotechnology, 20*, 592–596.

Sohni, A., & Verfaillie, C. M. (2013). Mesenchymal stem cells migration homing and tracking. *Stem Cells International, 2013*, 130763.

Soltysova, A., Altanerova, V., & Altaner, C. (2005). Cancer stem cells. *Neoplasma, 52*, 435–440.

Sordi, V., Malosio, M. L., Marchesi, F., Mercalli, A., Melzi, R., Giordano, T., Belmonte, N., Ferrari, G., Leone, B. E., Bertuzzi, F., Zerbini, G., Allavena, P., Bonifacio, E., & Piemonti, L. (2005). Bone marrow mesenchymal stem cells express a restricted set of functionally active chemokine receptors capable of promoting migration to pancreatic islets. *Blood, 106*, 419–427.

Spaeth, E. L., Dembinski, J. L., Sasser, A. K., Watson, K., Klopp, A., Hall, B., Andreeff, M., & Marini, F. (2009). Mesenchymal stem cell transition to tumor-associated fibroblasts contributes to fibrovascular network expansion and tumor progression. *PloS One, 4*, e4992.

Studeny, M., Marini, F. C., Champlin, R. E., Zompetta, C., Fidler, I. J., & Andreeff, M. (2002). Bone marrow-derived mesenchymal stem cells as vehicles for interferon-beta delivery into tumors. *Cancer Research, 62*, 3603–3608.

Studeny, M., Marini, F. C., Dembinski, J. L., Zompetta, C., Cabreira-Hansen, M., Bekele, B. N., Champlin, R. E., & Andreeff, M. (2004). Mesenchymal stem cells: Potential precursors for tumor stroma and targeted-delivery vehicles for anticancer agents. *Journal of the National Cancer Institute, 96*, 1593–1603.

Sun, X. Y., Nong, J., Qin, K., Warnock, G. L., & Dai, L. J. (2011). Mesenchymal stem cell-mediated cancer therapy: A dual-targeted strategy of personalized medicine. *World Journal of Stem Cells, 3*, 96–103.

Suzuki, K., Sun, R., Origuchi, M., Kanehira, M., Takahata, T., Itoh, J., Umezawa, A., Kijima, H., Fukuda, S., & Saijo, Y. (2011). Mesenchymal stromal cells promote tumor growth through the enhancement of neovascularization. *Molecular Medicine, 17*, 579–587.

Terskikh, A. V., Bryant, P. J., & Schwartz, P. H. (2006). Mammalian stem cells. *Pediatric Research, 59*, 13R–20R.

Teven, C. M., Liu, X., Hu, N., Tang, N., Kim, S. H., Huang, E., Yang, K., Li, M., Gao, J. L., Liu, H., Natale, R. B., Luther, G., Luo, Q., Wang, L., Rames, R., Bi, Y., Luo, J., Luu, H. H., Haydon, R. C., Reid, R. R., & He, T. C. (2011). Epigenetic regulation of mesenchymal stem cells: A focus on osteogenic and adipogenic differentiation. *Stem Cells International, 2011*, 201371.

Tolar, J., Nauta, A. J., Osborn, M. J., Panoskaltsis Mortari, A., Mcelmurry, R. T., Bell, S., Xia, L., Zhou, N., Riddle, M., Schroeder, T. M., Westendorf, J. J., Mcivor, R. S., Hogendoorn, P. C., Szuhai, K., Oseth, L., Hirsch, B., Yant, S. R., Kay, M. A., Peister, A., Prockop, D. J., Fibbe, W. E., & Blazar, B. R. (2007). Sarcoma derived from cultured mesenchymal stem cells. *Stem Cells, 25*, 371–379.

Tsai, M. S., Lee, J. L., Chang, Y. J., & Hwang, S. M. (2004). Isolation of human multipotent mesenchymal stem cells from second-trimester amniotic fluid using a novel two-stage culture protocol. *Human Reproduction, 19*, 1450–1456.

Tsai, K. S., Yang, S. H., Lei, Y. P., Tsai, C. C., Chen, H. W., Hsu, C. Y., Chen, L. L., Wang, H. W., Miller, S. A., Chiou, S. H., Hung, M. C., & Hung, S. C. (2011). Mesenchymal stem cells promote formation of colorectal tumors in mice. *Gastroenterology, 141*, 1046–1056.

Van't Hof, W., Mal, N., Huang, Y., Zhang, M., Popovic, Z., Forudi, F., Deans, R., & Penn, M. S. (2007). Direct delivery of syngeneic and allogeneic large-scale expanded multipotent adult progenitor cells improves cardiac function after myocardial infarct. *Cytotherapy, 9*, 477–487.

Vellasamy, S., Sandrasaigaran, P., Vidyadaran, S., George, E., & Ramasamy, R. (2012). Isolation and characterisation of mesenchymal stem cells derived from human placenta tissue. *World Journal of Stem Cells, 4*, 53–61.

Villaron, E. M., Almeida, J., Lopez-Holgado, N., Alcoceba, M., Sanchez-Abarca, L. I., Sanchez-Guijo, F. M., Alberca, M., Perez-Simon, J. A., San Miguel, J. F., & Del Canizo, M. C. (2004). Mesenchymal stem cells are present in peripheral blood and can engraft after allogeneic hematopoietic stem cell transplantation. *Haematologica, 89*, 1421–1427.

Wang, Y., Huso, D. L., Harrington, J., Kellner, J., Jeong, D. K., Turney, J., & Mcniece, I. K. (2005). Outgrowth of a transformed cell population derived from normal human BM mesenchymal stem cell culture. *Cytotherapy, 7*, 509–519.

Wynn, R. F., Hart, C. A., Corradi-Perini, C., O'neill, L., Evans, C. A., Wraith, J. E., Fairbairn, L. J., & Bellantuono, I. (2004). A small proportion of mesenchymal stem cells strongly expresses functionally active CXCR4 receptor capable of promoting migration to bone marrow. *Blood, 104*, 2643–2645.

Wyrsch, A., Dalle Carbonare, V., Jansen, W., Chklovskaia, E., Nissen, C., Surbek, D., Holzgreve, W., Tichelli, A., & Wodnar-Filipowicz, A. (1999).

Umbilical cord blood from preterm human fetuses is rich in committed and primitive hematopoietic progenitors with high proliferative and self-renewal capacity. *Experimental Hematology, 27*, 1338–1345.

Yagi, H., & Kitagawa, Y. (2013). The role of mesenchymal stem cells in cancer development. *Frontiers in Genetics, 4*, 261.

Yang, C., Lei, D., Ouyang, W., Ren, J., Li, H., Hu, J., & Huang, S. (2014). Conditioned media from human adipose tissue-derived mesenchymal stem cells and umbilical cord-derived mesenchymal stem cells efficiently induced the apoptosis and differentiation in human glioma cell lines in vitro. *BioMed Research International, 2014*, 109389.

Ye, H., Cheng, J., Tang, Y., Liu, Z., Xu, C., Liu, Y., & Sun, Y. (2012). Human bone marrow-derived mesenchymal stem cells produced TGFbeta contributes to progression and metastasis of prostate cancer. *Cancer Investigation, 30*, 513–518.

Yu, B., Zhang, X., & Li, X. (2014). Exosomes derived from mesenchymal stem cells. *International Journal of Molecular Sciences, 15*, 4142–4157.

Zeng, H. L., Zhong, Q., Qin, Y. L., Bu, Q. Q., Han, X. A., Jia, H. T., & Liu, H. W. (2011). Hypoxia-mimetic agents inhibit proliferation and alter the morphology of human umbilical cord-derived mesenchymal stem cells. *BMC Cell Biology, 12*, 32.

Zhou, Y. F., Bosch-Marce, M., Okuyama, H., Krishnamachary, B., Kimura, H., Zhang, L., Huso, D. L., & Semenza, G. L. (2006). Spontaneous transformation of cultured mouse bone marrow-derived stromal cells. *Cancer Research, 66*, 10849–10854.

Zhu, W., Xu, W., Jiang, R., Qian, H., Chen, M., Hu, J., Cao, W., Han, C., & Chen, Y. (2006). Mesenchymal stem cells derived from bone marrow favor tumor cell growth in vivo. *Experimental and Molecular Pathology, 80*, 267–274.

Zischek, C., Niess, H., Ischenko, I., Conrad, C., Huss, R., Jauch, K. W., Nelson, P. J., & Bruns, C. (2009). Targeting tumor stroma using engineered mesenchymal stem cells reduces the growth of pancreatic carcinoma. *Annals of Surgery, 250*, 747–753.

Zuk, P. A., Zhu, M., Ashjian, P., De Ugarte, D. A., Huang, J. I., Mizuno, H., Alfonso, Z. C., Fraser, J. K., Benhaim, P., & Hedrick, M. H. (2002). Human adipose tissue is a source of multipotent stem cells. *Molecular Biology of the Cell, 13*, 4279–4295.

Advs Exp. Medicine, Biology - Innovations in Cancer Research and Regenerative Medicine (2018) 1: 63–75
DOI 10.1007/5584_2017_71

Published online: 8 July 2017

Dental Pulp Stem Cells and Neurogenesis

Ibrahim Mortada, Rola Mortada, and Mohamad Al Bazzal

Abstract

Recent advances in regenerative medicine and cell-based therapy are bringing promising perspectives for the use of stem cells in clinical trials. Stem cells are undifferentiated cells capable of multilineage differentiation and available in numerous sources in the human body. Dental pulp constitutes an attractive source of these cells since collecting mesenchymal stem cells from this site is a noninvasive procedure which can be done following a common surgical extraction of supernumerary or wisdom teeth. Thus tissue sacrifice is very low and several cytotypes can be obtained owing to these cells' multipotency, in addition to the fact that they can be cryopreserved and stored for long periods. Mesenchymal stem cells have high proliferation rates making them favorable for clinical application. These multipotent cells present in a biological waste constitute an appropriate support in the management of many neurological disorders. After a brief overview on the different types of dental stem cells, this chapter will focus on the characteristics of dental pulp stem cells, their handling and applications in neural tissue engineering, as well as neural induction protocols leading to their potential therapeutic use in the management of neurological diseases.

Keywords

Biology • DPSCs • Experimental medicine • Regenerative medicine • Stem cells

I. Mortada (✉)
Faculty of Medicine, American University of Beirut, Beirut, Lebanon
e-mail: ikm03@aub.edu.lb

R. Mortada and M. Al Bazzal
Lebanese University School of Dentistry, Beirut, Lebanon

Abbreviations

a-MEM	Minimum essential medium, alpha modification
ATRA	All-trans retinoic acid
BMMSCs	Bone marrow mesenchymal stem cells
DFPCs	Dental follicle progenitor cells
DFSCs	Dental follicle stem cells
DMEM	Dulbecco's modified Eagle's medium
DMSO	Dimethyl sulfoxide
DSCs	Dental stem cells
ECM	Extracellular matrix
EGF	Epidermal growth factor
MAP 2	Microtubule-associated protein 2
MSCs	Mesenchymal stem cells
NSE	Neuron-specific enolase
PDLSCs	Periodontal ligament stem cells
rMSCs	Rat bone marrow mesenchymal stem cells
SCAP	Stem cells from apical papilla
SHED	Stem cells from human exfoliated deciduous teeth
TNC	Tenascin C

1 Introduction

In recent years, regenerative medicine has evolved rapidly with the discovery of multiple sources of mesenchymal stem cells (MSCs) (Hemmat et al. 2010). After the bone marrow and the adipose tissue, the oral cavity constitutes an important source of MSCs that exist in various areas such as the periodontal ligament, the dental follicle and mainly the dental pulp (Xiao and Nasu 2014). Dental stem cells (DSCs) are mainly characterized by their ease of availability since they are extracted from erupting primary teeth or extracted teeth. This makes their isolation much simpler and less invasive than aspiration of bone marrow mesenchymal stem cells (BMMSCs), for instance. Some studies even demonstrated the DSCs express-specific pluripotency markers (MYC, SOX2) which are absent in other mesenchymal stem cells (Yalvac et al. 2010). The absence of ethical considerations related to their use when compared to the use of totipotent and pluripotent cells and the absence of any history of cancer formation after their transplantation increase the interest in the DSCs.

The dental pulp constitutes the best source of oral stem cells. Under specific conditions, dental pulp stem cells (DPSCs) are capable of multilineage differentiation, among which is the neural differentiation (Gronthos et al. 2000). Extensive research focusing on DPSC-derived neural lineages showed that these cells have a characteristic innate neurogenic potential since they are originally issued from the embryonic neural crest (La Noce et al. 2014). The ability of DPSCs to generate neurons and glial cells makes them an exciting source in the management of neurodegenerative diseases. For this review, the focus is on the DPSCs, their characteristics, isolation, and their neural induction protocols.

2 Dental Stem Cells (DSCs)

As previously mentioned, teeth constitute a promising source of multipotent stem cells. Different populations of MSCs reside in teeth and they are nominated according to their harvest site (Racz et al. 2014; Ding et al. 2015). They present some variations related to their differentiation potential, cell surface receptors, and proliferative capacity although they share the same elongated fibroblast-like morphology that is characteristic to their mesenchymal character (Ding et al. 2015; Xiao and Nasu 2014; Zhao and Chai 2015). Figure 1 highlights the dental and associated tissues from which different populations of DSCs can be isolated.

Several types of DSCs were described.

2.1 Dental Pulp Stem Cells (DPSCs)

DPSCs were initially isolated from wisdom teeth by Gronthos et al. They were reported to have a high proliferation capacity and the potential to

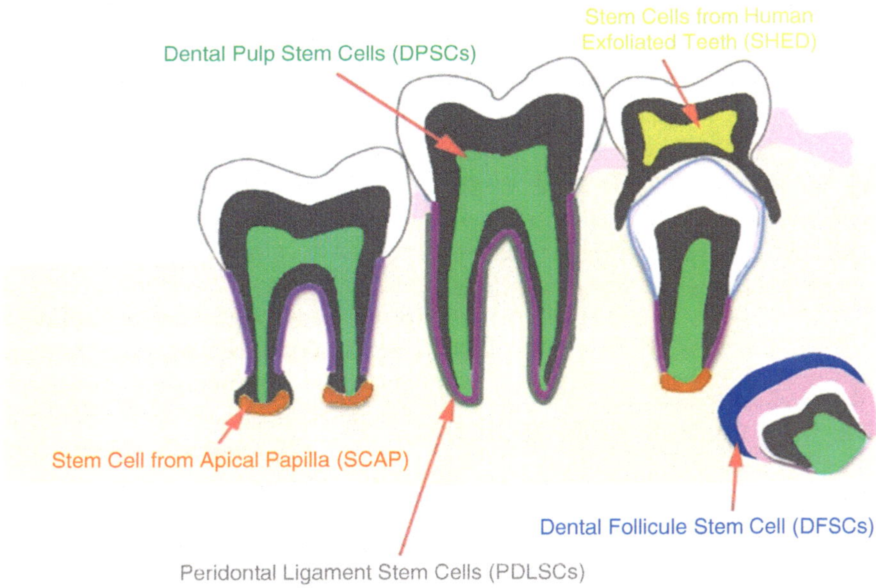

Fig. 1 Tissues of origin of the different populations of DSCs. DPSCs can be extracted from the inner tooth pulp of adult molars (*green*); SHED can be extracted from the pulp of deciduous exfoliated teeth (*yellow*); SCAP can be extracted from the apical papilla (*orange*); PDLSCs can be extracted from the periodontal ligament (*gray*); DFSCs can be extracted from the dental follicle (*blue*)

produce mineralized colonies (Gronthos et al. 2002). DPSCs are located in the dental crown, specifically in the pulp chamber (Baume 1980). The existing cells belong to heterogeneous populations of fibroblasts, osteoprogenitors neural, vascular, and immune cells (Goldberg and Smith 2004). The dental pulp has a high regenerative capacity and helps in the periodontal tissue repair (Dimitrova-Nakov et al. 2014; About 2013).

2.2 Stem Cells from Human Exfoliated Deciduous Teeth (SHED)

These progenitor cells were obtained for the first time from deciduous incisors in 2003. They express various embryonic markers (NANOG and OCT4) and MSCs markers (CD146 and STRO-1) (Sakai et al. 2012). Interestingly, they showed higher capability for differentiation and more proliferation rate than BMMSCs and even DPSCs. They can differentiate into different cell types, such as adipocytes, neural cells, and endothelial cells (Tatullo et al. 2015).

2.3 Periodontal Ligament Stem Cells (PDLSCs)

PDLSCs present similar characteristics to MSCs although they are derived from the neural crest cells (Chai et al. 2000). Those residing in the perivascular wall have similar cell morphology, differentiation potentials, and phenotype (Iwasaki et al. 2013). They resemble BMMSCs in their immunomodulatory ability and can differentiate into osteoblast, cementoblasts, and adipocytes (Wada et al. 2009).

2.4 Stem Cells from Apical Papilla (SCAP)

During dental root formation, SCAP are isolated from the apex. They have similar characteristics to MSCs and can be induced into chondrocytes, adipocytes, and neurons (Guo et al. 2013).

2.5 Dental Follicle Stem Cells (DFSCs)

During the early stages of tooth formation, DFSCs are isolated from dental follicle surrounding tooth germ (Silvério et al. 2012). The dental follicle is an ectomesenchymal condensation of cells. They can differentiate into neurons, osteoblasts, and chondrocytes (Saito et al. 2015).

3 Dental Pulp Stem Cells (DPSCs)

3.1 Dental Pulp as a Source of Multipotent Cells

Teeth are divided into two distinct anatomical entities, the crown and the root, connected to the supporting bone by the periodontal ligament. The crown comprises ameloblasts-generated enamel and odontoblasts-generated dentin in addition to the pulp. After tooth eruption, enamel formation stops occurring naturally due to the disappearance of ameloblasts from the surface. On the other hand, odontoblasts, which are present inside the pulp facing the dentin's inner surface, continue to deposit secondary dentin throughout life. They also form reparative or tertiary dentin in response to various chemical, mechanical, or bacterial stimuli. Interestingly, odontoblasts damage doesn't prohibit tertiary dentin formation in the presence of pulpal disruption. This reparative dentinogenesis was considered to be orchestrated by newly formed odontoblasts emerging from the pulp. Such findings were the starting point to the exploration of stem cell niches that might exist inside the dental pulp (Sonoyama et al. 2007).

DPSC were the first type of isolated DSCs (Gronthos et al. 2000).The pulp is a soft tissue enclosed within mineralized structures, the enamel, the dentin, and the cementum (Nanci 2007). It contains blood vessels, lymphatics, connective tissue, neural fibers (Liu et al. 2006), as well as DPSCs that continuously divide and undergo differentiation into different cell types (Gronthos et al. 2000). Functionally, the pulp detects unhealthy stimuli and assures the homeostasis of the tooth organ (Nanci 2007). It can be divided into four different layers. The external layer contains odontoblasts, the second layer contains mainly collagen fibers, and the third layer comprises undifferentiated cells and progenitor cells including the DPSCs. Undifferentiated cells migrate from this layer to various areas of the body where they can differentiate under specific conditions and various stimulations to give rise to specific cells. The deepest layer of the pulp comprises nerves and blood vessels (Goldberg and Smith 2004).

Dental pulp tissues from third molars, supernumerary teeth, or deciduous teeth constitute an easily accessible origin of MSCs without causing any morbidity to the patient as these teeth are often discarded. DPSCs express embryonic markers of stem cells and display multipotency markers indicating their possible spontaneous neural differentiation. They have higher proliferation rates and are more readily available than BMMSCs. Their multipotential capability and their distinct plasticity are not surprising since these cells come from both mesenchymal and ectodermic components, and they contain cells derived from the neural crest (Alge et al. 2010; D'aquino et al. 2009). The negative immunoreactivity for CD45 antibodies confirms that this cell line is not hematopoietic (D'aquino et al. 2007).

3.2 Neural Markers Expression by DPSCs

Foudah et al. confirmed that DPSCs spontaneously express the neural markers Nestin, NeuN, and β-III-tubulin (Foudah et al. 2014). Feng et al. demonstrated that they spontaneously express Nestin, β-III-tubulin, tyrosine hydroxylase, and microtubule-associated protein 2 (MAP2 (Feng et al. 2013). Govindasamy et al. confirmed that DPSCs exhibit higher spontaneous expression of the neural markers than SHED (PAX6 and Nestin) (Govindasamy et al. 2010). Martens et al. reported that DPSCs express synaptophysin, β-III-tubulin, and S100 protein with the presence of an eventual positive immunoreaction for galactocerebroside and NGF receptor p75 (Martens et al. 2012). Tamaki et al. confirmed that these undifferentiated mesenchymal stem cells exhibit a positive immune reactivity to antibodies against some neural markers like neurofilament (NF)-200, Nestin, and class β-III-tubulin (Tamaki et al. 2013). Karaoz et al. verified that DPSCs spontaneously express several neural markers associated with both neural stem cells and mature neural lineages of which NES, tenascin C (TNC), GFAP, connexin-43, MAP2ab, c-FOS, NEF-H, NEF-L and TUBB3, SOX2, ENO2 (Karaoz et al. 2010).

The amount of serum used in the medium seems to have an influence on the level of expression of neural markers in undifferentiated DPSCs. A high level of Nestin expression and an absence of β-III-tubulin and S100 are associated with a low serum level of around 2%. However, in high levels of serum of 10%, Nestin expression is relatively reduced (Ranganathan and Lakshminarayanan 2012).

On the other hand, tetrodotoxin, a potent neurotoxin, is quite specific in blocking the sodium ion channel and therefore the flow of sodium ions while having no effect on potassium ions. Therefore it blocks the conduction of nerve impulses along nerve fibers and axons. The patch-clamp recording technique detected a voltage-gated tetrodotoxin-sensitive inward current in undifferentiated DPSCs, which suggests a neural-like sodium conductance. This finding underlines the ability of these cells to exhibit some functions related to mature neurons (Davidson 1994).

4 Handling of Dental Pulp Stem Cells

4.1 Identification

Several techniques are used to identify adult stem cells:

- Identifying their specific surface markers
- Labeling the cells in the living organ with markers and then determining the differentiated cell types to which they give rise
- Removing stem cells from a living animal, labeling them, and transplanting them into a different animal to find out if they would migrate to their original tissue
- Isolating somatic stem cells, culturing them (adding growth factors, introducing genes)
- Infecting adult stem cells with a virus to identify them among others
- Testing their clonogenicity which means their "stemness" capability

4.2 Isolation

The isolation of DPSCs is usually achieved following two methods: the explants method and the enzymatic digestion of pulp tissue method. In the first method, the pulp is surgically removed and the cells are grown from tissue fragments (Roozafzoon et al. 2015), while in the second method, collagenase and dispase digest the dental pulp (Sun et al. 2014; Paschalidis et al. 2014), after which the cells are seeded. Cell proliferation is observed and the MSCs are characterized using flow cytometry based on staining with specific markers (Raoof et al. 2014). Furthermore, other studies noted that the isolation of more immature stem cells necessitates tissue explants in a multistage process in which the progenitor cells are first grown in culture, then enzymatically digested, followed by the expansion of isolated cells (Jung et al. 2012).

In the original protocol of DPSC isolation using impacted wisdom teeth as the harvesting source (Jung et al. 2012), the pulp was extirpated from the tooth, then an enzymatic digestion was performed. The obtained cells formed more colonies and had a higher rate of proliferation when compared to BMMSCs. Since then, numerous studies have explored alternative techniques and improved methods for isolation and culture of DPSCs (Kawashima 2012). Hilkens et al. (Hilkens et al. 2013) compared the enzymatic digestion and the cell outgrowth methods of isolation of stem cells from wisdom teeth. Results revealed no significant difference in differentiation potentials and markers expression. To date, a specific technique which is superior in terms of karyotypic stability, proliferative ability, or therapeutic use of DPSCs couldn't be reached; however, enzymatic digestion is more frequently employed than outgrowth explants (Ledesma-Martínez et al. 2016).

The type of tooth employed as a donor site of pulp tissue to be isolated was extensively investigated. Impacted third molars are frequently used, although exfoliated deciduous teeth are considered another excellent source of stem cells. Kerkis and Caplan (2012) isolated three distinct stem cell populations from deciduous teeth pulp. Govindasamy et al. (2011) reported higher proliferation rates and higher pluripotent and neuroectodermal markers in the cells isolated from deciduous teeth when compared to those issued from permanent teeth. Isolation of DPSCs derived from supernumerary teeth was also reported in the literature (Huang et al. 2008).

On the other hand, composition of the culture medium is of particular importance especially with the increasing tendency for clinical application of DPSC technologies (Ferro et al. 2012). For instance, bovine serum presents an eventual risk of bovine spongiforme encephalopathy if used in clinical therapy. In order to reduce this risk, a chemically defined medium with decreased levels of human serum can be used (Lizier et al. 2012).

4.3 Culture and Expansion

During their initial growth, self-renewal capacity of MSCs leads to colonies propagating from a unique cell on a plastic surface. These cells have an increased potential of expansion; therefore, the colonies formed in the first culture can be subcultured by several passages (Bakopoulou et al. 2011). Studies reported that DPSCs can be cultured for extended periods of time (6 months) without affecting their markers, their plasticity, or their morphology (Suchanek et al. 2010). In culture, cells were adherent and elongated in shape, with thin expansions (see Fig. 2). After reaching the maximum number of passages prior to entering senescence, these cells still had a normal karyotype and doubling period at up to 40 doublings. The capacity of DPSCs to be expanded to high numbers suggests an important

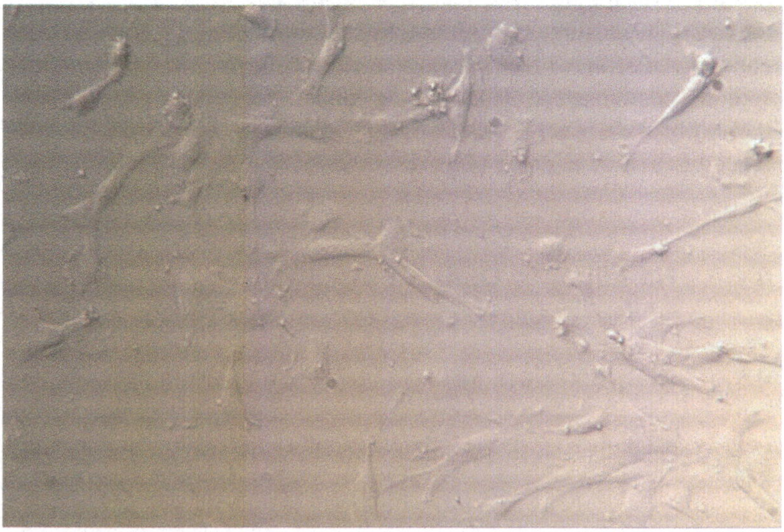

Fig. 2 Expanded cells in culture at day 15. They are spindle-shaped and comprise a homogenous cell population when viewed under the phase-contrast microscope (400×)

advantage for the use of these cells in several therapeutic applications (Laino et al. 2005).

Stem cells are seeded in specific conditions for the purpose of inducing their differentiation into the desired cell type. Growth factors, cell adhesion molecules, and other molecular signals are involved in the initiation and the supporting of the differentiation process.

4.4 Cryopreservation

Cryopreservation allows teeth banking and storage of DPSCs for future use. Cryopreserving cells for clinical use is a very important idea to consider. Studies showed no changes in markers and cell viability when using controlled cooling rates for cryopreservation (Lee et al. 2011). The use of liquid nitrogen allows cells to be cooled. Freezing in the presence of cryopreservative, such as dimethyl sulfoxide (DMSO), is required to inhibit the ice formation inside and around the cells and to prevent cell dehydration that would naturally lead to cell damage and death.

5 In Vitro Applications of DPSCs in Neural Tissue Engineering

5.1 Regulation of Differentiation

DPSCs exhibit a transcriptional change giving rise to progenitor cells without any obvious changes in self-renewal ability. This progression is the first step in the commitment process. At this stage, the progenitor cell has a more limited developmental program. The exit of tooth-derived MSCs from stemness into commitment occurs when the progenitor cell continues its division and acquires distinguishable features of fully committed mature cells with characteristic phenotypes. The differentiation of tooth-derived MSCs and their commitment to mature cell types, for example, neurons, is a well-established process that involves influence of chemical stimuli (forskolin, β-mercaptoethanol) and the activities of cytokines, transcription factors, extracellular matrix molecules, and growth factors (Doi et al. 2004).

5.2 Role of Chemicals

5.2.1 All-Trans Retinoic Acid (ATRA)

ATRA is a preinduction agent used before adding growth factors and a potent differentiation agent in various cells such as neuronal cells. It induces pluripotent cells differentiation into specific lineages among which are neural cell types in a time- and concentration-dependent manner (Kadar et al. 2009).

5.2.2 Dimethyl Sulfoxide (DMSO)

DMSO is a cell-differentiating agent and cryoprotectant. It is also used in the management of brain edema, amyloidosis, and schizophrenia (Santos et al. 2003). DMSO has a major role in the neural differentiation of stem cells, although the mechanism is still unknown. Compared to ATRA, DMSO has lower price and faster action; thus it is recommended as a preinduction factor for the neural differentiation of DPSCs.

5.3 Role of the Extracellular Matrix (ECM)

Differentiation is a well-controlled process highly influenced by the interaction between the cell and its microenvironment. This interaction constitutes the main element that will help conserve the stability of commitment genes. The ECM is in the heart of the cell microenvironment. Its main components are proteoglycans, glycosaminoglycans, and glycoproteins (Blau and Baltimore 1991). One type of interaction between ECM and growth factors is the binding of the basic amino acids of growth factors to heparan sulfate, the negatively charged chain of proteoglycans. Another example of interactions is the important role played by growth factors to regulate the production of the matrix proteins and their respective receptors leading to an appropriate gene expression regulation (Frescaline et al. 2013).

5.4 Role of Growth Factors

Neurotrophins have an important role in stimulating stem cell neural differentiation. It is therefore necessary to assess the specific role of each of these factors. Stable neural differentiation of stem cells can be achieved by combining chemicals with some growth factors, including NGF, bFGF, PDGF, and BDNF. Osathanon et al. investigated the possible implication of Notch signaling during DPSC neural differentiation induced by growth factor protocol (Osathanon et al. 2014). The authors detected a remarkable increase in mRNA expression level of Notch signaling target gene, *HEY1*, in growth factor-induced DPSC-derived neuronal-like cells. In fact, it was previously demonstrated that Notch signaling target genes, *HEY* family, regulates neural differentiation (Jalali et al. 2011; Mukhopadhyay et al. 2009). In addition, Notch signaling was found to be involved in neurogenic differentiation of PDLSCs (Osathanon et al. 2013). Furthermore, it was found that transfection with Notch intracellular domain and subsequent neurotrophic factors administration was associated with differentiation of mesenchymal stromal cells into neuronal cells (Dezawa et al. 2004). Therefore, current evidence implies the participation of Notch signaling pathway in growth factor-derived DPSC neural differentiation; however, further studies are warranted to elucidate the molecular mechanism of Notch signaling which regulates DPSC fate determination and neurogenic commitment.

We will briefly discuss the characteristics of the neurotrophins.

NGF is an influential factor and has a major role in the development and preservation of the sensory and sympathetic nervous systems. It promotes in vivo and in vitro neurites outgrowth, as well as nerve cell recovery following surgical, ischemic, or chemical injuries. By supporting survival and growth of neural cells, it is able to promote neural differentiation and regulate cell growth. In clinical therapy, NGF exhibits the ability to heal nerve injury. It has an antiapoptotic role in premature neurons and protects axons and myelin from inflammatory damage in order to modulate the immune system. NGF was also found to induce BMMSC differentiation into neural cells, through the generation of neuropeptide signals and receptors. Therefore, NGF is essential for stem cell neural differentiation, and eventually, it could help in treating injured nerves (Aloe et al. 2015; Hu et al. 2016).

bFGF mediates cell proliferation and mitosis and intensifies spinal cord repair and neuronal axon regeneration via the expression of nerve-specific proteins. It is expressed in both embryonic and adult central and peripheral nervous systems, promotes sympathetic and parasympathetic nervous axon growth, maintains neuronal cell survival, and induces damaged nerve repair and neurites outgrowth. Moreover, bFGF can promote differentiation of adrenal pheochromocytoma cells-12. It is a potent growth factor in the neural differentiation of DPSCs by DPSCs-stimulating colony-forming units and upregulation of embryonic stem cell markers Oct4, Nanog, and Rex-1. Increased DPSC neurosphere size and upregulation of neurogenic markers were seen in the presence of bFGF. Hu et al. showed that pretreatment with bFGF promotes neural specification (Hu et al. 2016).

PDGF-BB is an important mitogen for connective tissue cells and is implicated in hyperplasia and embryonic neuron development. PDGF promotes stem cell proliferation and protects them against senescence, apoptosis, and immunomodulatory defects (Hata et al. 2010).

BDNF promotes neuronal differentiation, survival during early development, neural plasticity, and adult neurogenesis. It stimulates and controls oligodendrocyte progenitor cells development in vitro and in vivo (Tsiperson et al. 2015). This signaling molecule regulates neuronal plasticity, proliferation, cell growth, and survival. The combination of stem cells with BDNF promotes differentiation of primitive cells originating from the bone marrow into the glial lineage. Park et al. used a retroviral vector carrying rats' BDNF cDNA to transduce rat bone marrow mesenchymal stem cells (rMSCs). Following their intravitreal or subretinal injection, these rMSCs were incorporated into the retina, with further production of BDNF (Park et al. 2012).

6 DPSCs: Overview of Protocols for Neural Induction

6.1 Preinduction

DPSCs vary in their differentiation potential. For instance, subpopulations that express the p75 neurotrophin receptor are considered to have superior neural differentiation ability (Dai et al. 2013). To date, a wide variety of protocols leading to in vitro neural differentiation of DPSCs are described in the literature. Many variations exist among these protocols related to the culture medium (basal medium, growth factors, and supplements) but also to the employed surface coating, the culture duration, as well as the existence of a single or multiple culture stages.

Spontaneous neural induction demonstrated by the expression of neural markers was reported in the literature (Gervois et al. 2015; Osathanon et al. 2014). Xiao et al. found that DPSCs cultured in suspension under serum-free conditions could spontaneously differentiate into the neural cells (Xiao and Tsutsui 2013).

6.2 Multistage Culture

Currently, in vitro DPSC neurogenic induction protocols are shifting toward multistage and more complex culture protocols. These are considered to more accurately reproduce the dynamic environment of DPSC niche during neural differentiation. Pretreatment with β-mercaptoethanol was shown to enhance neural progenitors' survival in addition to promoting Nestin expression as shown by Ni et al. (Ni et al. 2001). Lu et al. noted that a 5-h exposure to β-mercaptoethanol (5 mM of concentration) led to the expression of neuron-specific enolase (NSE) and NF-200 by multipotent adipose-derived cells (Lu et al. 2012). 5-azacytidine is a demethylation agent that would consequently erase the epigenetic memory and make DPSCs more pliable to neural lineage differentiation. This agent was also used for the stem cells' preinduction treatment before their exposition to neural cultures (Kiraly et al. 2011).

6.3 Surface Coating

The use of surface coating for cell culture is another variable in the in vitro protocols, and it has an important influence on the neurogenic outcome. The culture of neurosphere suspension occurs in nonadherent culture dishes (Gervois et al. 2015). However, most of the adherent culture occurs on bare tissue culture-treated polystyrene (TCPS). A considerable number of studies reported the culture of DPSCs on different surface coatings such as polyornithine with laminin (Gervois et al. 2015), collagen type IV (Gervois et al. 2015; Lu et al. 2012) and collagen type I (Van Kooten et al. 2004), and chitosan (Osathanon et al. 2014).

Importantly, these surface coatings were chosen for DPSC neural induction based on trial and error of commonly utilized surface coatings for cell culture (Feng et al. 2014). To date, there is no systematic comparison of the various surface coatings used for the neural induction of DPSCs.

6.4 Culture Medium (Basal Medium, Growth Factors, Culture Supplements, and Small Molecules)

The great majority of studies used either Dulbecco's modified Eagle's medium (DMEM)/F12 medium or the neurobasal medium. Few studies used the Eagle's minimum essential medium, alpha modification (a-MEM).

Brewer et al. (Brewer et al. 1993) developed the neurobasal medium. Its formulation was optimized for the survival of in vitro cultured neural cell types. This improvement is best exemplified by its lower osmolarity compared to DMEM and the decrease of glutamine and cysteine concentrations in addition to the removal of toxic ferrous sulfate present in DMEM/F12. An extensive variation exists among the growth factors utilized in the neurogenic induction culture medium. The basic fibroblast growth factor (bFGF) and the epidermal growth factor (EGF) are the two most commonly supplemented growth factors. Research has shown that both bFGF and EGF are potent in stimulating the proliferation of neural stem cells (Santa-Olalla and Covarrubias 1995). The concentrations of bFGF and EGF used for DPSC neurogenic differentiation range from 10 to 50 mg/ml. The last component of the neural induction medium includes a large diversity of small molecules. They include small molecules without any neural inductive effects such as antibiotics (penicillin, streptomycin), antimycotics (amphotericin B), energy supplements (l-glutamine), forskolin, and vitamins such as ascorbic acid. Many combinations of these molecules were utilized in the various published neural inductive protocols of DPSCs, although the optimal combination for inducing neurogenesis is still currently unclear (Kim et al. 2011).

7 Conclusion

The concluding remarks, conceived after a thorough review of published literature, point to an important role for DPSCs in neurogenesis and the advancement of regenerative medicine. The selection of appropriate healthy teeth extracted from young patients with a good medical history, the adoption of ameliorated techniques of stem cell isolation and culture, as well as the employment of the best neural induction protocols constitute the mainstay of a very promising revolution in this field. However, the long-term side effects associated with the use of DPSCs and their potential to transform into tumors over time have not been sufficiently studied thus far. More studies are warranted to clarify possible long-term risks related to the use of these cells, as well as their interactions with the immune system of the host.

Conflicts of Interest The authors declare no conflicts of interest in relation to this article.

References

About, I. (2013). Dentin–pulp regeneration: The primordial role of the microenvironment and its modification by traumatic injuries and bioactive materials. *Endodontic Topics, 28*, 61–89.

Alge, D. L., Zhou, D., Adams, L. L., Wyss, B. K., Shadday, M. D., Woods, E. J., Chu, T. M. G., & Goebel, W. S. (2010). Donor-matched comparison of dental pulp stem cells and bone marrow-derived mesenchymal stem cells in a rat model. *Journal of Tissue Engineering and Regenerative Medicine, 4*, 73–81.

Aloe, L., Rocco, M. L., Omar Balzamino, B., & Micera, A. (2015). Nerve growth factor: A focus on Neuroscience and therapy. *Current Neuropharmacology, 13*, 294–303.

Bakopoulou, A., Leyhausen, G., Volk, J., Tsiftsoglou, A., Garefis, P., Koidis, P., & Geurtsen, W. (2011). Assessment of the impact of two different isolation methods on the osteo/odontogenic differentiation potential of human dental stem cells derived from deciduous teeth. *Calcified Tissue International, 88*, 130–141.

Baume, L. J. (1980). The biology of pulp and dentine. A historic, terminologic-taxonomic, histologic-biochemical, embryonic and clinical survey. *Monographs in Oral Science, 8*, 1–220.

Blau, H. M., & Baltimore, D. (1991). Differentiation requires continuous regulation. *The Journal of Cell Biology, 112*, 781–783.

Brewer, G. J., Torricelli, J. R., Evege, E. K., & Price, P. J. (1993). Optimized survival of hippocampal neurons in B27-supplemented Neurobasal, a new serum-free medium combination. *Journal of Neuroscience Research, 35*, 567–576.

Chai, Y., Jiang, X., Ito, Y., Bringas, P., Jr., Han, J., Rowitch, D. H., Soriano, P., Mcmahon, A. P., & Sucov, H. M. (2000). Fate of the mammalian cranial neural crest during tooth and mandibular morphogenesis. *Development, 127*, 1671–1679.

Dai, J. W., Yuan, H., Shen, S. Y., Lu, J. T., Zhu, X. F., Yang, T., Zhang, J. F., & Shen, G. F. (2013). p75 neurotrophin receptor positive dental pulp stem cells: New hope for patients with neurodegenerative disease and neural injury. *Shanghai Kou Qiang Yi Xue, 22*, 469–472.

D'aquino, R., Graziano, A., Sampaolesi, M., Laino, G., Pirozzi, G., De Rosa, A., & Papaccio, G. (2007). Human postnatal dental pulp cells co-differentiate into osteoblasts and endotheliocytes: A pivotal synergy leading to adult bone tissue formation. *Cell Death and Differentiation, 14*, 1162–1171.

D'aquino, R., De Rosa, A., Laino, G., Caruso, F., Guida, L., Rullo, R., Checchi, V., Laino, L., Tirino, V., & Papaccio, G. (2009). Human dental pulp stem cells: From biology to clinical applications. *Journal of Experimental Zoology. Part B, Molecular and Developmental Evolution, 312b*, 408–415.

Davidson, R. M. (1994). Neural form of voltage-dependent sodium current in human cultured dental pulp cells. *Archives of Oral Biology, 39*, 613–620.

Dezawa, M., Kanno, H., Hoshino, M., Cho, H., Matsumoto, N., Itokazu, Y., Tajima, N., Yamada, H., Sawada, H., Ishikawa, H., Mimura, T., Kitada, M., Suzuki, Y., & Ide, C. (2004). Specific induction of neuronal cells from bone marrow stromal cells and application for autologous transplantation. *Journal of Clinical Investigation, 113*, 1701–1710.

Dimitrova-Nakov, S., Baudry, A., Harichane, Y., Kellermann, O., & Goldberg, M. (2014). Pulp stem cells: Implication in reparative dentin formation. *Journal of Endodontia, 40*, S13–S18.

Ding, G., Niu, J., & Wei, F. (2015). Current understanding of orofacial tissue derived mesenchymal stem cells: An immunological perspective. *Histology and Histopathology, 30*, 255–265.

Doi, M., Nagano, A., & Nakamura, Y. (2004). Molecular cloning and characterization of a novel gene, EMILIN-5, and its possible involvement in skeletal development. *Biochemical and Biophysical Research Communications, 313*, 888–893.

Feng, X., Xing, J., Feng, G., Sang, A., Shen, B., Xu, Y., Jiang, J., Liu, S., Tan, W., Gu, Z., & Li, L. (2013). Age-dependent impaired neurogenic differentiation capacity of dental stem cell is associated with Wnt/beta-catenin signaling. *Cellular and Molecular Neurobiology, 33*, 1023–1031.

Feng, X., Lu, X., Huang, D., Xing, J., Feng, G., Jin, G., Yi, X., Li, L., Lu, Y., Nie, D., Chen, X., Zhang, L., Gu, Z., & Zhang, X. (2014). 3D porous chitosan scaffolds suit survival and neural differentiation of dental pulp stem cells. *Cellular and Molecular Neurobiology, 34*, 859–870.

Ferro, F., Spelat, R., Beltrami, A. P., Cesselli, D., & Curcio, F. (2012). Isolation and characterization of human dental pulp derived stem cells by using media containing low human serum percentage as clinical grade substitutes for bovine serum. *PloS One, 7*, e48945.

Foudah, D., Monfrini, M., Donzelli, E., Niada, S., Brini, A. T., Orciani, M., Tredici, G., & Miloso, M. (2014). Expression of neural markers by undifferentiated mesenchymal-like stem cells from different sources. *Journal of Immunology Research, 2014*, 16.

Frescaline, G., Bouderlique, T., Mansoor, L., Carpentier, G., Baroukh, B., Sineriz, F., Trouillas, M., Saffar, J.-L., Courty, J., Lataillade, J.-J., Papy-Garcia, D., & Albanese, P. (2013). Glycosaminoglycan mimetic associated to human mesenchymal stem cell-based scaffolds inhibit ectopic bone formation, but induce angiogenesis in vivo. *Tissue Engineering. Part A, 19*, 1641–1653.

Gervois, P., Struys, T., Hilkens, P., Bronckaers, A., Ratajczak, J., Politis, C., Brône, B., Lambrichts, I., & Martens, W. (2015). Neurogenic maturation of human dental pulp stem cells following Neurosphere generation induces morphological and electrophysiological characteristics of functional neurons. *Stem Cells and Development, 24*, 296–311.

Goldberg, M., & Smith, A. J. (2004). Cells and extracellular matrices of dentin and pulp: A biological basis for repair and tissue engineering. *Critical Reviews in Oral Biology and Medicine, 15*, 13–27.

Govindasamy, V., Abdullah, A. N., Ronald, V. S., Musa, S., Ab Aziz, Z. A., Zain, R. B., Totey, S., Bhonde, R. R., & Abu Kasim, N. H. (2010). Inherent differential propensity of dental pulp stem cells derived from human deciduous and permanent teeth. *Journal of Endodontia, 36*, 1504–1515.

Govindasamy, V., Ronald, V. S., Abdullah, A. N., Ganesan Nathan, K. R., Aziz, Z. A., Abdullah, M., Zain, R. B., Kasim, N. H., Musa, S., & Bhonde, R. R. (2011). Human platelet lysate permits scale-up of dental pulp stromal cells for clinical applications. *Cytotherapy, 13*, 1221–1233.

Gronthos, S., Mankani, M., Brahim, J., Robey, P. G., & Shi, S. (2000). Postnatal human dental pulp stem cells (DPSCs) in vitro and in vivo. *Proceedings of the National Academy of Sciences of the United States of America, 97*, 13625–13630.

Gronthos, S., Brahim, J., Li, W., Fisher, L. W., Cherman, N., Boyde, A., Denbesten, P., Robey, P. G., & Shi, S. (2002). Stem cell properties of human dental pulp stem cells. *Journal of Dental Research, 81*, 531–535.

Guo, L., Li, J., Qiao, X., Yu, M., Tang, W., Wang, H., Guo, W., & Tian, W. (2013). Comparison of odontogenic differentiation of human dental follicle cells and human dental papilla cells. *PloS One, 8*, e62332.

Hata, N., Shinojima, N., Gumin, J., Yong, R., Marini, F., Andreeff, M., & Lang, F. F. (2010). PDGF-BB mediates the tropism of human mesenchymal stem cells for malignant gliomas. *Neurosurgery, 66*, 144–157.

Hemmat, S., Lieberman, D. M., & Most, S. P. (2010). An introduction to stem cell biology. *Facial Plastic Surgery, 26*, 343–349.

Hilkens, P., Gervois, P., Fanton, Y., Vanormelingen, J., Martens, W., Struys, T., Politis, C., Lambrichts, I., & Bronckaers, A. (2013). Effect of isolation methodology on stem cell properties and multilineage differentiation potential of human dental pulp stem cells. *Cell and Tissue Research, 353*, 65–78.

Hu, Y., Zhang, Y. A. N., Tian, K., Xun, C., Wang, S., & Lv, D. (2016). Effects of nerve growth factor and basic fibroblast growth factor dual gene modification on rat bone marrow mesenchymal stem cell differentiation into neuron-like cells in vitro. *Molecular Medicine Reports, 13*, 49–58.

Huang, A. H., Chen, Y. K., Lin, L. M., Shieh, T. Y., & Chan, A. W. (2008). Isolation and characterization of dental pulp stem cells from a supernumerary tooth. *Journal of Oral Pathology & Medicine, 37*, 571–574.

Iwasaki, K., Komaki, M., Yokoyama, N., Tanaka, Y., Taki, A., Kimura, Y., Takeda, M., Oda, S., Izumi, Y., & Morita, I. (2013). Periodontal ligament stem cells possess the characteristics of pericytes. *Journal of Periodontology, 84*, 1425–1433.

Jalali, A., Bassuk, A. G., Kan, L., Israsena, N., Mukhopadhyay, A., Mcguire, T., & Kessler, J. A. (2011). HeyL promotes neuronal differentiation of neural progenitor cells. *Journal of Neuroscience Research, 89*, 299–309.

Jung, S., Panchalingam, K. M., Rosenberg, L., & Behie, L. A. (2012). Ex vivo expansion of human mesenchymal stem cells in defined serum-free media. *Stem Cells International, 2012*, 21.

Kadar, K., Kiraly, M., Porcsalmy, B., Molnar, B., Racz, G. Z., Blazsek, J., Kallo, K., Szabo, E. L., Gera, I., Gerber, G., & Varga, G. (2009). Differentiation potential of stem cells from human dental origin - promise for tissue engineering. *Journal of Physiology and Pharmacology, 60*(Suppl 7), 167–175.

Karaoz, E., Dogan, B. N., Aksoy, A., Gacar, G., Akyuz, S., Ayhan, S., Genc, Z. S., Yuruker, S., Duruksu, G., Demircan, P. C., & Sariboyaci, A. E. (2010). Isolation and in vitro characterisation of dental pulp stem cells from natal teeth. *Histochemistry and Cell Biology, 133*, 95–112.

Kawashima, N. (2012). Characterisation of dental pulp stem cells: A new horizon for tissue regeneration? *Archives of Oral Biology, 57*, 1439–1458.

Kerkis, I., & Caplan, A. I. (2012). Stem cells in dental pulp of deciduous teeth. *Tissue Engineering. Part B, Reviews, 18*, 129–138.

Kim, H., Zahir, T., Tator, C. H., & Shoichet, M. S. (2011). Effects of dibutyryl cyclic-AMP on survival and neuronal differentiation of neural stem/progenitor cells transplanted into spinal cord injured rats. *PloS One, 6*, e21744.

Kiraly, M., Kadar, K., Horvathy, D. B., Nardai, P., Racz, G. Z., Lacza, Z., Varga, G., & Gerber, G. (2011). Integration of neuronally predifferentiated human dental pulp stem cells into rat brain in vivo. *Neurochemistry International, 59*, 371–381.

La Noce, M., Mele, L., Tirino, V., Paino, F., De Rosa, A., Naddeo, P., Papagerakis, P., Papaccio, G., & Desiderio, V. (2014). Neural crest stem cell population in craniomaxillofacial development and tissue repair. *European Cells & Materials, 28*, 348–357.

Laino, G., D'aquino, R., Graziano, A., Lanza, V., Carinci, F., Naro, F., Pirozzi, G., & Papaccio, G. (2005). A new population of human adult dental pulp stem cells: A useful source of living autologous fibrous bone tissue (LAB). *Journal of Bone and Mineral Research, 20*, 1394–1402.

Ledesma-Martínez, E., Mendoza-Núñez, V. M., & Santiago-Osorio, E. (2016). Mesenchymal stem cells derived from dental pulp: A review. *Stem Cells International, 2016*, 12.

Lee, J. H., Lee, D. S., Choung, H. W., Shon, W. J., Seo, B. M., Lee, E. H., Cho, J. Y., & Park, J. C. (2011). Odontogenic differentiation of human dental pulp stem cells induced by preameloblast-derived factors. *Biomaterials, 32*, 9696–9706.

Liu, H., Gronthos, S., & Shi, S. (2006). Dental pulp stem cells. *Methods in Enzymology, 419*, 99–113.

Lizier, N. F., Kerkis, A., Gomes, C. M., Hebling, J., Oliveira, C. F., Caplan, A. I., & Kerkis, I. (2012). Scaling-up of dental pulp stem cells isolated from multiple niches. *PloS One, 7*, e39885.

Lu, Y., Yuan, X., Ou, Y., Cai, Y., Wang, S., Sun, Q., & Zhang, W. (2012). Autophagy and apoptosis during adult adipose-derived stromal cells differentiation into neuron-like cells in vitro. *Neural Regeneration Research, 7*, 1205–1212.

Martens, W., Wolfs, E., Struys, T., Politis, C., Bronckaers, A., & Lambrichts, I. (2012). Expression pattern of basal markers in human dental pulp stem cells and tissue. *Cells, Tissues, Organs, 196*, 490–500.

Mukhopadhyay, A., Jarrett, J., Chlon, T., & Kessler, J. A. (2009). HeyL regulates the number of TrkC neurons in dorsal root ganglia. *Developmental Biology, 334*, 142–151.

Nanci, A. (2007). *Ten Cate's oral histology - Pageburst on VitalSource: Development, structure, and function.* Elsevier Health Sciences.

Ni, L., Wen, Y., Peng, X., & Jonakait, G. M. (2001). Antioxidants N-acetylcysteine (NAC) and 2-mercaptoethanol (2-ME) affect the survival and differentiative potential of cholinergic precursors from the embryonic septal nuclei and basal forebrain: Involvement of ras signaling. *Brain Research. Developmental Brain Research, 130*, 207–216.

Osathanon, T., Manokawinchoke, J., Nowwarote, N., Aguilar, P., Palaga, T., & Pavasant, P. (2013). Notch signaling is involved in neurogenic commitment of human periodontal ligament-derived mesenchymal stem cells. *Stem Cells and Development, 22*, 1220–1231.

Osathanon, T., Sawangmake, C., Nowwarote, N., & Pavasant, P. (2014). Neurogenic differentiation of human dental pulp stem cells using different induction protocols. *Oral Diseases, 20*, 352–358.

Park, H. Y., Kim, J. H., Sun Kim, H., & Park, C. K. (2012). Stem cell-based delivery of brain-derived neurotrophic factor gene in the rat retina. *Brain Research, 1469*, 10–23.

Paschalidis, T., Bakopoulou, A., Papa, P., Leyhausen, G., Geurtsen, W., & Koidis, P. (2014). Dental pulp stem cells' secretome enhances pulp repair processes and compensates TEGDMA-induced cytotoxicity. *Dental Materials, 30*, e405–e418.

Racz, G. Z., Kadar, K., Foldes, A., Kallo, K., Perczel-Kovach, K., Keremi, B., Nagy, A., & Varga, G. (2014). Immunomodulatory and potential therapeutic role of mesenchymal stem cells in periodontitis. *Journal of Physiology and Pharmacology, 65*, 327–339.

Ranganathan, K., & Lakshminarayanan, V. (2012). Stem cells of the dental pulp. *Indian Journal of Dental Research, 23*, 558.

Raoof, M., Yaghoobi, M. M., Derakhshani, A., Kamal-Abadi, A. M., Ebrahimi, B., Abbasnejad, M., & Shokouhinejad, N. (2014). A modified efficient method for dental pulp stem cell isolation. *Dent Res J (Isfahan), 11*, 244–250.

Roozafzoon, R., Lashay, A., Vasei, M., AI, J., Khoshzaban, A., Keshel, S. H., Barabadi, Z., & Bahrami, H. (2015). Dental pulp stem cells differentiation into retinal ganglion-like cells in a three dimensional network. *Biochemical and Biophysical Research Communications, 457*, 154–160.

Saito, M. T., Silvério, K. G., Casati, M. Z., Sallum, E. A., & Nociti, F. H., Jr. (2015). Tooth-derived stem cells: Update and perspectives. *World Journal of Stem Cells, 7*, 399–407.

Sakai, K., Yamamoto, A., Matsubara, K., Nakamura, S., Naruse, M., Yamagata, M., Sakamoto, K., tauchi, R., wakao, N., Imagama, S., Hibi, H., Kadomatsu, K., Ishiguro, N., & Ueda, M. (2012). Human dental pulp-derived stem cells promote locomotor recovery after complete transection of the rat spinal cord by multiple neuro-regenerative mechanisms. *The Journal of Clinical Investigation, 122*, 80–90.

Santa-Olalla, J., & Covarrubias, L. (1995). Epidermal growth factor (EGF), transforming growth factor-alpha (TGF-alpha), and basic fibroblast growth factor (bFGF) differentially influence neural precursor cells of mouse embryonic mesencephalon. *Journal of Neuroscience Research, 42*, 172–183.

Santos, N. C., Figueira-Coelho, J., Martins-Silva, J., & Saldanha, C. (2003). Multidisciplinary utilization of dimethyl sulfoxide: Pharmacological, cellular, and molecular aspects. *Biochemical Pharmacology, 65*, 1035–1041.

Silvério, K. G., Davidson, K. C., James, R. G., Adams, A. M., Foster, B. L., Nociti, F. H., Somermam, M. J., & Moon, R. T. (2012). Wnt/β-catenin pathway regulates Bmp2-mediated differentiation of dental follicle cells. *Journal of Periodontal Research, 47*. doi:10.1111/j.1600-0765.2011.01433.x.

Sonoyama, W., Yamaza, T., Gronthos, S., & Shi, S. (2007). Multipotent stem cells in dental pulp. In *Culture of human stem cells*. Hoboken: Wiley.

Suchanek, J., Visek, B., Soukup, T., El-Din Mohamed, S. K., Ivancaková, R., Mokrỳ, J., Aboul-Ezz, E. H., & Omran, A. (2010). Stem cells from human exfoliated deciduous teeth--isolation, long term cultivation and phenotypical analysis. *Acta Medica (Hradec Králové), 53*, 93–99.

Sun, H.-H., Chen, B., Zhu, Q.-L., Kong, H., Li, Q.-H., Gao, L.-N., Xiao, M., Chen, F.-M., & Yu, Q. (2014). Investigation of dental pulp stem cells isolated from discarded human teeth extracted due to aggressive periodontitis. *Biomaterials, 35*, 9459–9472.

Tamaki, Y., Nakahara, T., Ishikawa, H., & Sato, S. (2013). In vitro analysis of mesenchymal stem cells derived from human teeth and bone marrow. *Odontology, 101*, 121–132.

Tatullo, M., Marrelli, M., Shakesheff, K. M., & White, L. J. (2015). Dental pulp stem cells: Function, isolation and applications in regenerative medicine. *Journal of Tissue Engineering and Regenerative Medicine, 9*, 1205–1216.

Tsiperson, V., Huang, Y., Bagayogo, I., Song, Y., Vondran, M. W., Dicicco-Bloom, E. & Dreyfus, C. F. 2015. Brain-derived neurotrophic factor deficiency restricts proliferation of oligodendrocyte progenitors following Cuprizone-induced demyelination. ASN Neuro, 7, 1759091414566878.

Van Kooten, T. G., Spijker, H. T., & Busscher, H. J. (2004). Plasma-treated polystyrene surfaces: Model surfaces for studying cell-biomaterial interactions. *Biomaterials, 25*, 1735–1747.

Wada, N., Menicanin, D., Shi, S., Bartold, P. M., & Gronthos, S. (2009). Immunomodulatory properties of human periodontal ligament stem cells. *Journal of Cellular Physiology, 219*, 667–676.

Xiao, L., & Nasu, M. (2014). From regenerative dentistry to regenerative medicine: Progress, challenges, and potential applications of oral stem cells. *Stem Cells and Cloning: Advances and Applications, 7*, 89–99.

Xiao, L., & Tsutsui, T. (2013). Characterization of human dental pulp cells-derived spheroids in serum-free medium: Stem cells in the core. *Journal of Cellular Biochemistry, 114*, 2624–2636.

Yalvac, M. E., Ramazanoglu, M., Rizvanov, A. A., Sahin, F., Bayrak, O. F., Salli, U., Palotas, A., & Kose, G. T. (2010). Isolation and characterization of stem cells derived from human third molar tooth germs of young adults: Implications in neo-vascularization, osteo-, adipo- and neurogenesis. *The Pharmacogenomics Journal, 10*, 105–113.

Zhao, H., & Chai, Y. (2015). Stem cells in teeth and craniofacial bones. *Journal of Dental Research, 94*, 1495–1501.

Advs Exp. Medicine, Biology - Innovations in Cancer Research and Regenerative Medicine (2018) 1: 77–87
DOI 10.1007/5584_2017_117

Published online: 4 January 2018

Single-Cell Expression Profiling and Proteomics of Primordial Germ Cells, Spermatogonial Stem Cells, Adult Germ Stem Cells, and Oocytes

Sabine Conrad, Hossein Azizi, and Thomas Skutella

Abstract

The mammalian germ cells, cell assemblies, tissues, and organs during development and maturation have been extensively studied at the tissue level. However, to investigate and understand the fundamental insights at the molecular basis of germ and stem cells, their cell fate plasticity, and determination, it is of most importance to analyze at the large scale on the single-cell level through different biological windows. Here, modern molecular techniques optimized for single-cell analysis, including single fluorescence-activated cell sorting (FACS) and single-cell RNA sequencing (scRNA-seq) or microfluidic high-throughput quantitative real-time polymerase chain reaction (qRT-PCR) for single-cell gene expression and liquid chromatography coupled to tandem mass spectrometry (LC-MSMS) for protein profiling, have been established and are still getting optimized.

This review aims on describing and discussing recent single-cell expression profiling and proteomics of different types of human germ cells, including primordial germ cells (PGCs), spermatogonial stem cells (SSCs), human adult germ stem cells (haGSCs), and oocytes.

Keywords

Germ cells • Single-cell expression profiling • Single-cell proteomics

S. Conrad
P.O. Box 1243, 72072, Tübingen, Germany
e-mail: saco_de@yahoo.de

H. Azizi
Faculty of Biotechnology, Amol University of Special Modern Technologies, Amol, Iran
e-mail: hosseinazizi1358@gmail.com

T. Skutella (✉)
Institute for Anatomy and Cell Biology III, Medical Faculty, Heidelberg University, Heidelberg, Germany
e-mail: skutella@ana.uni-heidelberg.de

Abbreviations

FACS	Fluorescence-activated cell sorting
haGSC	Human adult germ stem cell
hESC	Human embryonic stem cells
hFib	Human fibroblast
hPSC	Human pluripotent stem cell
hSSC	Human spermatogonial stem cell
LC-MSMS	Liquid chromatography mass spectrometry
MACS	Magnetic activated cell sorting
MSCs	Mesenchymal stem cells
hPGC	Human primordial germ cell
qRT-PCR	Quantitative real-time polymerase chain reaction
scRNA-seq	Single-cell RNA sequencing
hSSC	Human spermatogonial stem cell

1 Introduction into Single-Cell Analysis

The biology of single cells and its variability can be examined with advanced technologies in cell sciences like the newly developed single-cell genomic and proteomic platforms (Huang et al. 2015). They can be employed for genome-wide scale analyses that link pheno- and genotypes at cellular and subcellular levels at different cell biological stages or conditions (Gerovska and Arauzo-Bravo 2016; Grindberg et al. 2013; Nakamura et al. 2015; Saliba et al. 2014; Shapiro et al. 2013; Tang et al. 2011). Based on single-cell transcriptional profiling and characterization, cells can not only be utilized to display accurate images of cellular stages but also for bulk studies of embryogenesis, maturation, and pathological conditions or regeneration.

Important information can be achieved by single-cell transcriptomic and proteomic analysis concerning cell populations and their variability, for example, uniform gene expressions in one population or high variability and its dynamic variations with time. The analysis of a cell pool or tissue versus single cells provides only average expression values, while certain important information on variability and differential cell regulation is completely lost. Since all cell systems are built up of cellular heterogeneity with diverging expression profiles and weak and strong signals, and even more in in vivo systems, the development of techniques to study heterogeneity mechanisms is essential. Single-cell analysis can be justified, when there are heterogeneities, except strong signals outstand over background noise. Thus, various bifurcations and stable points could be found. Bulk data for finding weak signals are also subject to masking effects, which are related to the timing of the event provoking signals. At the beginning of certain processes, often weak signals that are permanently amended during these progresses are sent out by single cells. Studying bulk samples for the examination of a dynamic event, its regulation on the transcriptomic and proteomic level might not be found, since it requires a sufficient amplitude differing from the background. Contradictory, a sooner discovery of an event could be achieved when utilizing single-cell measurements in a related context, because the background does not mask weaker event-related signals.

In recent times, scRNAseq and computational analysis tools gave access to the whole transcriptome and thus supply a useful tool for research studies to investigate transcription factors more consistently (Tanay and Regev 2017; Wagner et al. 2016). Due to single-cell technology in particular, developmental and germ cell biologists can now perform non-masked and concomitant assays of cell stage, lineage relationships, and genes controlling the underlying processes. Various factors influence the characteristics of each cell stage. While several factors are only temporary, like those during the cell cycle stages, others have long-lasting responsibilities, like transcription factor networks and signaling cascades in stem cell development (Graf and Enver 2009; Novershtern et al. 2011).

Consequently, the important questions can be raised in which temporary, long-lasting, or final cell stages are processes of interest. The transitional phases between those stages should be examined, as well as the factors that trigger these passages.

In accordance with former studies about cell differentiation, single-cell analysis results also demonstrate that the quiescence of specific transcription factors is triggered selectively (Graf and Enver 2009; Novershtern et al. 2011). Moreover, it is reported that in the developmental stage, a "single-pulse" pattern can be found in transcriptional profiles (Yosef and Regev 2011).

Single-cell studies on different stages of embryonic development provided insight in transcription factor networks involved in development and differentiation (Choi et al. 2012; Giritharan et al. 2010; Maekawa et al. 2007). In mouse embryos, nine cell stages were detected by Jang et al. when employing this method. They first examined ESCs during differentiation into progenitor cells for various germ layers, and, second, they arranged the determined nine stages in a lineage tree (Jang et al. 2017). The research followed the task to examine the system for influence of transcriptional regulation on the development of cells. The employed flow cytometry and live-cell microscopy, to observe several biomarkers for cells stages, revealed the relative stability of cell stages, whereas the passage from one stage to another took place rather quickly. This outcome was the cause to concentrate more on cell stages than on their passages. Specifically, Jang et al. investigated thoroughly for interdependencies between certain transcription factors that could adjust into several stages of stability, which again could depict a certain cell stage (Huang et al. 2009). They projected and proved that, during the epiblast-like stage, the Oct4 expression is strongly influenced by Sox2 overexpression, whereas during the embryonic stem cell-like stage, it is less affected.

Insights into the single-cell chromatin stage were quite helpful for the categorization into cell stages and for the identification of highly active transcription factors (Tanay and Regev 2017; Wagner et al. 2016). The transition between different cell stages can be directly monitored by lineage tracking technologies (Woodworth et al. 2017).

The self-renewing, pluripotent and lineage primed stages of human pluripotent stem cells (hPSCs) are clearly characterized by single-cell expression profiles (Hough et al. 2014). For a clear distinction on the single-cell level, these cell subpopulations need to be analyzed via single-cell gene expression by Quantitative real-time polymerase chain reaction (qRT-PCR) with Fluidigm microfluidics system. This system provided a practicable solution for parallel high-throughput analysis in nanoliter volume wells. Growth factors and their receptors and transcription factors were encoded by the examined gene panel, which all played a role in pluripotency maintenance; this gene roster also contained transcription factors that trigger an early specification into somatic and extraembryonic lineages. The surface antigen expression clarified the single-cell gene expression analysis in cell subsets, which made obvious that hESC cultures abound as constant cell stages, actually under certain conditions inducing self-renewal. Canonical pluripotency transcription factors were expressed by large population parts, which could also transform into derivatives of all three germ layers. Smaller cell subpopulation parts were highly capable of self-renewing, as they continually showed no lineage priming and high transcripts of all examined pluripotency-related genes. The findings of Hough et al. suggest an inherent heterogeneity of pluripotent cells, which can readily become specified into extraembryonic or somatic lineages. Although epigenetic or genetic factors induce sustained self-renewal and repress differentiation, hPSCs can only depart from the pluripotency state undergoing successions of intermediary stages with the specific characteristic of lineage priming, instead of transforming to a new cell stage by quantum transition. These outcomes signal as well the difficulty of distinguishing naïve or primed pluripotency stages solely based on population data (Hough et al. 2014).

1.1 Human Primordial Germ Cells

In 2015, Guo et al. analyzed for the first time the transcriptome and DNA methylome landscapes of hPGCs at a single-cell and single-base resolution (Guo et al. 2015). The single-cell sequencing analysis displayed that specific transcription patterns were displayed by PGCs that include a synchronous expression of pluripotency genes as well as germline-specific genes, while a subset of them showed characteristics that were specific for their developmental stage. The single-cell transcriptomes of hPGCs were relatively homogenous between the migrating and gonadal stage. But it became obvious that individual female PGCs exhibited strong heterogeneous gene expression patterns, which were not observed before. An interesting observation was also made in so far that hPGCs were found with strong enrichment of the base excision repair pathway, which is a part of the local DNA methylation processing PGCs (Ooi and Bestor 2008). Moreover, by examining the DNA methylome of hPGCs in parallel, Guo et al. reported a global demethylation of their genomes. About 10–11 weeks of post-gestation, the PGCs have almost no DNA methylation; the median methylation levels in male and female PGCs were only 7.8% and 6.0%, respectively. Thus, Guo et al. made it feasible to decipher the complex epigenetic germline reprogramming in order to restore the totipotency in fertilized oocytes. The expression of CD117 (also known as KIT) by human gonadal PGCs in general has been demonstrated at the surface area of the cell, although hPGCs do not express SOX2 (Perrett et al. 2008; Robinson et al. 2001). Lately, FACS was employed to isolate human PGCs in the gonads, so that their transcriptome at the 16-week gestation stage could be analyzed (Gkountela et al. 2013). Pluripotency marker genes such as POU5F1 (also known as OCT4), NANOG (Nanog homeobox), ZFP42 (also known as REX1), DPPA3 (also known as STELLA), SALL4 (spalt like transcription factor 4), and LIN28A (lin-28 homolog A) were clearly expressed by human mitotic PGCs. However, they did not express SOX2 (SRY (sex determining region Y)-box 2), as previously reported (Hayashi et al. 2011; Magnusdottir et al. 2013; Ohinata et al. 2009). The germline-specific marker genes like KIT, ALPL (also known as TNAP), TFAP2C (also known as AP2g), DND1 (DND microRNA-mediated repression inhibitor 1), NANOS3 (nanos C2HC-type zinc finger 3), and TCL1A (T-cell leukemia/lymphoma 1A) were also expressed by hPGCs.

Recently, Li et al. (2017) tried to identify the developmental trajectories and heterogeneity of fetal female germ cells (Li et al. 2017). The authors conducted scRNA-seq analysis of more than 2000 germ cells and their gonadal niche cells at several developmental stages. The main observations from this study include the description of distinct transcriptomic profiles of transcription factor networks at the different developmental stages. The single-cell expression profiling showed that female embryos contain at least four stages of fetal germ cells. Further, it could be observed that bidirectional bone morphogenetic protein (BMP) and notch signaling is present between female germ cells and their niche. This work presents a detailed roadmap for human germ cell development, which provides a guideline for the production of female germ cells from PSCs in vitro, and also helps to understand germ cell-related diseases.

1.2 Human Spermatogonial Stem Cells

von Kopylow et al. (2016) reported of the isolation and quantitative real-time PCR gene expression analysis of single potential human spermatogonial stem cells (hSSCs), which were defined by the combined expression of the cell surface protein fibroblast growth factor receptor 3 (FGFR3), a human undifferentiated type A spermatogonia marker, as well as the undifferentiated embryonic cell transcription factor 1 (UTF1), a spermatogonial nuclear marker (Kooistra et al. 2009; Kristensen et al. 2008; Okuda et al. 1998; Valli et al. 2014; von Kopylow et al. 2010). From 37 patients with full spermatogenesis and three patients with meiotic arrest, von Kopylow et al. utilized two 30 mg biopsies per patient to specifically select the FGFR3-positive spermatogonial subpopulation. In more detail, cell selection with magnetic beads in combination with a fluorescence-activated cell sorter antibody directed against human FGFR3 was applied to mark and visually detect the human FGFR3-positive spermatogonia. In a next step a micromanipulator was used to pick positively selected and bead-labeled cells and to avoid the contamination with more differentiated spermatogonia and of surrounding somatic cells. von Kopylow et al. found that the FGFR3 expression in the human testis is limited to a small subpopulation of nonproliferating, non-differentiating type A spermatogonia, which accumulate only in small cell clusters of two or four cells (von Kopylow et al. 2012a, b) and co-express FGFR3 and UTF1.

In conclusion, the authors consider that a subclassification of stemness based on transcript abundance is caused by the transcriptional heterogeneity of the SSC population. Thus, the development of new isolation protocols for selecting hSSCs, which show a high expression of specific stem cell markers found in the membrane, is of utmost importance.

Neuhaus et al. reported of single-cell gene expression analysis with RT-PCR that revealed diversity among human spermatogonia (Neuhaus et al. 2017). Further two single OCT4 positive cells from a patient with qualitatively normal spermatogenesis were selected for shallow RNA-seq.

Contrary to expectations, there were heterogeneous expression profiles found in the single-cell expression data from the patients with arrest of spermatogenesis (20 cells). Moreover, patients with normal spermatogenesis showed heterogeneous expression patterns of known undifferentiated as well as differentiated marker genes (OCT4, UTF1, and MAGE (MAGE family member A3) A4). The differentiated marker genes BOLL (boule homolog, RNA binding protein) and PRM2 (protamine 2) were found in each spermatogonia cluster (13 clusters with 85 cells). It was demonstrated that selected candidate markers (DDX5 (DEAD-box helicase 5), TSPY1 (testis specific protein, Y-linked 1), EEF1A1 (eukaryotic translation elongation factor 1 alpha 1), and NGN3 (neurogenin 3)) that are characteristic for spermatogonia have a heterogeneous protein expression. The BOLL and PRM2 expression was missing in the single-cell data from patients with arrest of spermatogenesis. The histological findings also conformed the lack of meiotic cells displayed (Grassetti et al. 2012; Luetjens et al. 2004).

Out of 105 selected cells that were analyzed for single-cell qRT-PCR expression, 101 cells displayed germ cell marker expression. Additionally, highly expressed spermatogonial markers could be identified by RNA-seq analysis of two individual cells, although a high heterogeneity of spermatogonia at the RNA and the protein level could be detected, instead of finding a specific molecular fingerprint.

Intriguingly, out of 105 single analyzed cells, only two (1.9%) expressed OCT4 in combination with germ cell-specific marker genes. The RNA-seq data generated from these two cells could not be used to compare differential gene expression from cell to cell, as coverage of the whole transcriptome was too low. However, the authors concluded that the OCT4 expression at the RNA level could prove distinct undifferentiated germ cell features, although they did not use primers for the stem cell-specific OCT4a. The authors did also not analyze OCT4 with antibody staining at the protein level, probably due to low abundance of the OCT4 protein in single human cells.

In many cases, Neuhaus et al. (2017) observed that cells with a differentiated profile expressed the meiotic and spermatid marker genes BOLL and PRM2, which is a characteristic of germ cell clusters, also expressed undifferentiated marker genes. Moreover, the expression variance of diverse cells from the same patient was higher than that of cells from different individual patients (Neuhaus et al. 2017).

In a recent study with neonatal mice, Hermann et al. also proved the transcriptional heterogeneity of spermatogonia (Hermann et al. 2015). Here, the microfluidic Fluidigm system for sc qRT-PCR analysis was utilized to examine 172 genes in spermatogonia and displayed present spermatogonia subtypes that also might function specifically (Hermann et al. 2015).

It is feasible that technical progress in the area of single-cell transcriptional analyses (Grindberg et al. 2013; Saliba et al. 2014) one day enables the identification of SSC markers yet unknown. On top of that, this approach could assist to expose the differential gene expression patterns involved with the symmetric or asymmetric divisions of human spermatogonia and moreover evaluate the cell response on characteristic growth factors or culture conditions in general. It is most important to define specific features of the hSSC system, thus determining further steps to find out the necessary conditions for efficient hSSC culture and in vitro spermatogenesis. Moreover, the underlying molecular mechanisms causing male infertility could be analyzed by single-cell approaches.

The reported snapshots from the studies of Kopylow et al. (2016) and Neuhaus et al. (2017) display the heterogeneity of human spermatogonia at the RNA and protein levels. For further analysis of the functional meaning of this heterogeneity and the stem cell populations' dynamics, it is necessary to design new approaches to simplify a repeated individual cell analysis. These outcomes imply that heterogeneous expression profiles could be a specific key for human spermatogonia, and they support the idea of a model heterogeneous stem cell population. Prospective research studies with fertile and infertile patients will help to evaluate the dynamics of spermatogonial populations.

1.3 Human Adult Germ Stem Cells

haGSCs and haGSC colonies that are developing out of CD49f magnetic activated cell sorting (MACS)- and the collagen-/laminin-binding matrix-selected fraction of enriched human adult spermatogonia were examined and reported by Conrad et al. (Conrad et al. 2016). In this research, a Fluidigm BioMark system for single-cell transcriptional profiling of long-term haGSCs was employed. When comparing human embryonic stem cells (hESCs) and human fibroblast cells (hFibs), hESCs showed their characteristic pluripotency profile. In contrast to hFibs, haGSCs revealed a characteristic germ- and pluripotency-associated gene expression profile. Genome-wide comparisons with microarray analysis revealed that haGSC colonies provide a heterogeneous gene expression and pluripotency profile and are in a greater or lesser extent unequal. Findings proved that in vitro, haGSCs have a particular molecular gene expression profile and are adult stem cells. Although they occur similar to genuine PSCs, they are not equivalent. At the molecular level, which reflects the cell plasticity and differentiation of haGSC colonies, they maintain their partial stage of pluripotency.

Single-cell expression profiling demonstrated that haGSCs are a heterogeneous cell population distinguishable from somatic cells such as hFibs or mesenchymal stem cells (MSCs), which do not display the typical germ and pluripotency expression. During cell culture, haGSCs were selected from matrix-selected spermatogonia and the enhanced CD49f MACS population. haGSCs can never be selected from patients without spermatogonia, like those suffering from sertoli-cell-only syndrome. In recurring experiments with sertoli-cell-only syndrome patients, the generation of haGSC colonies was virtually impossible. Due to the similar features of haGSC colonies and early hESC colonies, it was not difficult to find the morphological differences in haGSC colonies which typify a pivotal cluster with overgrowing epithelial cells. In primary cultures, first small haGSC colonies and islands developed 4–6 weeks after culturing enriched spermatogonia in hGSC mediums. Succeeding, denser haGSC aggregations were manually selected for further propagation and characterization. Characteristic for haGSC colonies was their surrounding of inner colony parts and exuberant epithelial cells that were close to early hESC colonies. Single-cell Fluidigm analysis revealed the substantial contrast between haGSCs and hFibs, regarding the germ cell- and pluripotency-associated gene expression. Merely in a few outliner hESCs and haGSCs, prevalent properties of hFibs could be detected; however, most of them had not many resemblances. Comparing hESC with haGSC showed an enhanced expression of pluripotency-related genes, like SOX2,

NANOG, LIN28, LIN28B, GDF3 (growth differentiation factor 3), CDH1 (cadherin 1), OCT4a, TDGF1 (teratocarcinoma-derived growth factor 1), and UTF1, by hESCs. Simultaneously, the expression of the germ cell-related genes CD9 (CD9 molecule), GFRA1 (GDNF family receptor alpha 1), NANOS1, STAT3 (signal transducer and activator of transcription 3), TSPYL, ADGRA3 (adhesion G protein-coupled receptor A3), and MYC (MYC proto-oncogene, bHLH transcription factor) by haGSCs was higher. The assessment of all haGSCs showed a similar profile with decreased expression levels, which means that haGSCs expressed basic pluripotency-related genes including OCT4a, NANOG, SOX2, and LIN28. In some measures, these genes still sustained their germ cell-related gene expression profile. The core pluripotency-related genes were less expressed in haGSCs as against hESCs, but they were substantially higher as against hFibs. Colonies proved to be heterogeneous and had similar properties like an ESC-like pluripotency stage. Moreover, in a microarray study, haGSC colonies were quite heterogeneous in their genes associated with expression of germ and pluripotency. As proven by Mizrak et al. (Mizrak et al. 2010), Chikovskaya et al. (Chikhovskaya et al. 2012), and Gonzalez et al. (Gonzalez et al. 2009), the separation of different testicular stem cell populations from tissue and their culture in vitro was practicable. Eventually, the existence of various stem cell types in adult human testicles is feasible.

In future single-cell studies with haGSCs, it may be interesting to analyze the release of the molecular block that inhibits haGSCs from fully converting to molecular PSCs in order to advance culture conditions.

1.4 Single-Cell Analysis of Human Oocytes

Our initial work about single human oocyte expression profiling with the BioMark scRT-PCR system (Fluidigm) demonstrated the capacity of this methodology to practice quantitative single-cell analyses with various genes being analyzed simultaneously (Virant-Klun et al. 2013). In the preceding study, Virant-Klun et al. analyzed 19 single human oocytes at various maturation stages (6 germinal vesicle (GV), 4 metaphase I (MI), 5 in vitro matured (IVM), and 4 mature metaphase II (MII) oocytes). After the pre-amplification procedure, the Fluidigm system was employed to analyze 56 genes related to PSCs and oocytes. The outcomes demonstrated that a small amount of only three outstanding oocytes (two immature MI oocytes and one in vitro matured oocyte IVM) were distinguishable from other oocytes at the molecular level. The heat map, the hierarchical clusters, and the principal component analysis showed that they did not cluster with other oocytes. No statistical differences in gene expression could be detected among them. Nevertheless, Virant-Klun et al. implied that large amounts of single oocytes at respective maturation stages would have to be analyzed in order to apply this methodology in the future.

Liu et al. (2016) provided a single-cell analysis in transcriptomic profiles of human oocytes and cumulus cells from different stages of oocyte development, including immature GV, MI, and MII cell stage from polycystic ovary syndrome (PCOS) patients. Transcriptional gene expression of oocytes and cumulus cells was determined via RNA sequencing. The authors could demonstrate that the sensitivity of the RNA sequencing technique is higher as it covered more data than conventional microarray methods. Different stages of oocytes, cumulus cells (CCs), and other heterogeneous single cells could be distinguished via single-cell RNA sequencing (Liu et al. 2016).

It has been observed that during assisted reproductive techniques, PCOS patients show higher amounts of proliferation and fertilization than those of non-PCOS patients and they produced the same amount of high-quality embryos, but the molecular basis was unclear.

Comprising, the meiosis maturation of PCOS oocytes is dysfunctional, as well as their gap junction, hormone response, DNA damaging, and secreted factors in the early GV phase. Thus, oocyte meiosis at the GV stage is delayed by malfunctioning genes, which may also hinder the fertilization and other processes. Possible causes for PCOS oocytes and CCs disorder at early stages could be identified after examining differentially expressed genes such as PPP2R1A (protein phosphatase 2 scaffold subunit Aalpha), PDGFRA (platelet derived growth factor receptor alpha), EGFR (epidermal growth factor receptor), GJA1 (gap junction protein alpha 1), PTGS2 (prostaglandin-endoperoxide synthase 2), TNFAIP6 (TNF alpha induced protein 6), TGF-β1 (latent transforming growth factor beta binding protein 1), CAV1 (caveolin 1), INHBB (inhibin beta B subunit), and others.

However, at MII stage induced by assisted reproductive techniques, these genes were normally expressed. These findings suggest that the quality of cumulus-oocyte complex (COC) can be enhanced by assisted reproductive techniques, which also augment fertilization and proliferation ratios in PCOS women.

In an attempt to better comprehend the composition and diversity of the proteome in the course of human oocyte maturation, Virant-Klun et al. (Virant-Klun et al. 2016) were able to generate initial proteome and also secretome maps of single human oocytes. In an initial attempt to establish the methodological platform for the single oocyte proteomic analysis, they employed a new serum-free hanging drop culture system to collect first 100 oocytes and finally detected 2.154 different proteins in this pooled cell population. Examining these proteins revealed that oocytes are mostly resting cells; it was suggested that their proteome is mainly customized for homeostasis, attachment to cells, and secretory factors (vesicle-mediated transport, extracellular region, and extracellular matrix proteins) to interact with the ovarian cells and tissue, while proteins involved in growth and proliferation, and processes like chromatin organization, transcription, splicing, and translation are under-expressed.

The gene ontology (GO) classes "sexual reproduction" and "fertilization" were distinguished to sort 32 oocyte-enriched proteins, which included ZP1–ZP4 (originating the zona pellucida), CD9 (important for sperm-egg recognition and fusion), BMP15 (bone morphogenetic protein 15) and GDF9 (oocyte-specific growth factors needed for follicle development), HENMT1 (methyltransferase stabilizing piRNAs, necessary for gametogenesis), BCL2L10 (apoptosis suppressor), CHEK1 (essential for cell cycle arrest), and PTTG1 (hindering segregation of chromosomes). The majority of the proteins are prominent for functioning in the development and maturation of gametes. Moreover, the oocyte-enriched proteins KHDC3L (KH domain containing 3 like, subcortical maternal complex member), PIWIL3 (piwi like RNA-mediated gene silencing 3), and NLRP7 (NLR family pyrin domain containing 7) could be detected, which were found lacking in high coverage proteomic studies about certain human cell lines and tissues.

A new technology for proteomic sample preparation with magnetic beads was employed to a novel protocol for sample preparation, SP3, which was introduced recently for low-input proteomics (Hughes et al. 2014), and thus the analysis of proteomes could be narrowed by successive approximation to the single-cell level. Although the protein content was low with merely ~100 ng per cell, ~450 proteins from single oocytes could be recognized consequently. The comparison of individual oocytes of GV and MII stage revealed that the tudor and KH domain-containing protein (TDRKH) is preferentially expressed in immature oocytes. However, Wee2 (WEE1 homolog 2), PCNA (proliferating cell nuclear antigen), and DNMT1 (DNA methyltransferase 1) were enriched in mature cells. Overall, this fact suggests that during oocyte maturation, sustaining the genome integrity is especially important.

This research proves that an innovative proteomic workflow makes single human oocyte analysis approachable, although only a relatively low number of ~450 proteins were detectable from single cells with the techniques applied.

Both studies in general prove that single human oocyte biology and preimplantation development can be further examined at the transcriptomic and proteomic level. Furthermore, the way for quantitative proteomics in other quantity-limited tissues and cell types might be established through this research study.

One drawback of this study is the quality of the oocytes obtained in the in vitro fertilization program. Oocytes included in this study would otherwise be discarded in daily medical practice, like immature (GV or MI) oocytes that cannot be fertilized and mature (MII) oocytes that did not fertilize in vitro.

2 Outlook

Future aspects might include the possibility to gather various kinds of single-cell data from all different kinds of female and male germ cells at different kinds of development and maturation stages. Together with an accessible integrative interpretation system for sharing these information, this will certainly produce new findings about cell development, conversion, and fate in the germline both in vivo and in vitro. Heterogeneity in single-cell expression profiling of human germ cells will definitely define important cellular subpopulations. Important lineage-specific genes, intracellular signaling molecules, and transcription factor networks will be classified to build up a more comprehensive molecular roadmap of the human germline both under normal and pathological conditions.

References

Chikhovskaya, J. V., Jonker, M. J., Meissner, A., Breit, T. M., Repping, S., & van Pelt, A. M. (2012). Human testis-derived embryonic stem cell-like cells are not pluripotent, but possess potential of mesenchymal progenitors. *Human Reproduction, 27*, 210–221.

Choi, I., Carey, T. S., Wilson, C. A., & Knott, J. G. (2012). Transcription factor AP-2gamma is a core regulator of tight junction biogenesis and cavity formation during mouse early embryogenesis. *Development, 139*, 4623–4632.

Conrad, S., Azizi, H., Hatami, M., Kubista, M., Bonin, M., Hennenlotter, J., Sievert, K. D., & Skutella, T. (2016). Expression of genes related to germ cell lineage and Pluripotency in single cells and colonies of human adult germ stem cells. *Stem Cells International, 2016*, 8582526.

Gerovska, D., & Arauzo-Bravo, M. J. (2016). Does mouse embryo primordial germ cell activation start before implantation as suggested by single-cell transcriptomics dynamics? *Molecular Human Reproduction, 22*, 208–225.

Giritharan, G., Li, M. W., Di Sebastiano, F., Esteban, F. J., Horcajadas, J. A., Lloyd, K. C., Donjacour, A., Maltepe, E., & Rinaudo, P. F. (2010). Effect of ICSI on gene expression and development of mouse preimplantation embryos. *Human Reproduction, 25*, 3012–3024.

Gkountela, S., Li, Z., Vincent, J. J., Zhang, K. X., Chen, A., Pellegrini, M., & Clark, A. T. (2013). The ontogeny of cKIT+ human primordial germ cells proves to be a resource for human germ line reprogramming, imprint erasure and in vitro differentiation. *Nature Cell Biology, 15*, 113–122.

Gonzalez, R., Griparic, L., Vargas, V., Burgee, K., Santacruz, P., Anderson, R., Schiewe, M., Silva, F., & Patel, A. (2009). A putative mesenchymal stem cells population isolated from adult human testes. *Biochemical and Biophysical Research Communications, 385*, 570–575.

Graf, T., & Enver, T. (2009). Forcing cells to change lineages. *Nature, 462*, 587–594.

Grassetti, D., Paoli, D., Gallo, M., D'Ambrosio, A., Lombardo, F., Lenzi, A., & Gandini, L. (2012). Protamine-1 and -2 polymorphisms and gene expression in male infertility: An Italian study. *Journal of Endocrinological Investigation, 35*, 882–888.

Grindberg, R. V., Yee-Greenbaum, J. L., McConnell, M. J., Novotny, M., O'Shaughnessy, A. L., Lambert, G. M., Arauzo-Bravo, M. J., Lee, J., Fishman, M., Robbins, G. E., et al. (2013). RNA-sequencing from single nuclei. *Proceedings of the National Academy of Sciences of the United States of America, 110*, 19802–19807.

Guo, F., Yan, L., Guo, H., Li, L., Hu, B., Zhao, Y., Yong, J., Hu, Y., Wang, X., Wei, Y., et al. (2015). The Transcriptome and DNA Methylome landscapes of human primordial germ cells. *Cell, 161*, 1437–1452.

Hayashi, K., Ohta, H., Kurimoto, K., Aramaki, S., & Saitou, M. (2011). Reconstitution of the mouse germ cell specification pathway in culture by pluripotent stem cells. *Cell, 146*, 519–532.

Hermann, B. P., Mutoji, K. N., Velte, E. K., Ko, D., Oatley, J. M., Geyer, C. B., & McCarrey, J. R. (2015). Transcriptional and translational heterogeneity among neonatal mouse spermatogonia. *Biology of Reproduction, 92*, 54.

Hough, S. R., Thornton, M., Mason, E., Mar, J. C., Wells, C. A., & Pera, M. F. (2014). Single-cell gene expression profiles define self-renewing, pluripotent, and lineage primed states of human pluripotent stem cells. *Stem Cell Reports, 2*, 881–895.

Huang, S., Ernberg, I., & Kauffman, S. (2009). Cancer attractors: A systems view of tumors from a gene network dynamics and developmental perspective. *Seminars in Cell & Developmental Biology, 20*, 869–876.

Huang, L., Ma, F., Chapman, A., Lu, S., & Xie, X. S. (2015). Single-cell whole-genome amplification and sequencing: Methodology and applications. *Annual Review of Genomics and Human Genetics, 16*, 79–102.

Hughes, C. S., Foehr, S., Garfield, D. A., Furlong, E. E., Steinmetz, L. M., & Krijgsveld, J. (2014). Ultrasensitive proteome analysis using paramagnetic bead technology. *Molecular systems biology 10*, 757.

Jang, S., Choubey, S., Furchtgott, L., Zou, L.N., Doyle, A., Menon, V., Loew, E.B., Krostag, A.R., Martinez, R.A., Madisen, L., et al. (2017). Dynamics of embryonic stem cell differentiation inferred from single-cell transcriptomics show a series of transitions through discrete cell states. *eLife, 6*.

Kooistra, S. M., Thummer, R. P., & Eggen, B. J. (2009). Characterization of human UTF1, a chromatin-associated protein with repressor activity expressed in pluripotent cells. *Stem Cell Research, 2*, 211–218.

Kristensen, D. M., Nielsen, J. E., Skakkebaek, N. E., Graem, N., Jacobsen, G. K., Rajpert-De Meyts, E., & Leffers, H. (2008). Presumed pluripotency markers UTF-1 and REX-1 are expressed in human adult testes and germ cell neoplasms. *Human Reproduction, 23*, 775–782.

Li, L., Dong, J., Yan, L., Yong, J., Liu, X., Hu, Y., Fan, X., Wu, X., Guo, H., Wang, X., et al. (2017). Single-cell RNA-Seq analysis maps development of human Germline cells and gonadal niche interactions. *Cell Stem Cell, 20*, 891–892.

Liu, Q., Li, Y., Feng, Y., Liu, C., Ma, J., Li, Y., Xiang, H., Ji, Y., Cao, Y., Tong, X., et al. (2016). Single-cell analysis of differences in transcriptomic profiles of oocytes and cumulus cells at GV, MI, MII stages from PCOS patients. *Scientific Reports, 6*, 39638.

Luetjens, C. M., Xu, E. Y., Rejo Pera, R. A., Kamischke, A., Nieschlag, E., & Gromoll, J. (2004). Association of meiotic arrest with lack of BOULE protein expression in infertile men. *The Journal of Clinical Endocrinology and Metabolism, 89*, 1926–1933.

Maekawa, M., Yamamoto, T., Kohno, M., Takeichi, M., & Nishida, E. (2007). Requirement for ERK MAP kinase in mouse preimplantation development. *Development, 134*, 2751–2759.

Magnusdottir, E., Dietmann, S., Murakami, K., Gunesdogan, U., Tang, F., Bao, S., Diamanti, E., Lao, K., Gottgens, B., & Azim Surani, M. (2013). A tripartite transcription factor network regulates primordial germ cell specification in mice. *Nature Cell Biology, 15*, 905–915.

Mizrak, S. C., Chikhovskaya, J. V., Sadri-Ardekani, H., van Daalen, S., Korver, C. M., Hovingh, S. E., Roepers-Gajadien, H. L., Raya, A., Fluiter, K., de Reijke, T. M., et al. (2010). Embryonic stem cell-like cells derived from adult human testis. *Human Reproduction, 25*, 158–167.

Nakamura, T., Yabuta, Y., Okamoto, I., Aramaki, S., Yokobayashi, S., Kurimoto, K., Sekiguchi, K., Nakagawa, M., Yamamoto, T., & Saitou, M. (2015). SC3-seq: A method for highly parallel and quantitative measurement of single-cell gene expression. *Nucleic Acids Research, 43*, e60.

Neuhaus, N., Yoon, J., Terwort, N., Kliesch, S., Seggewiss, J., Huge, A., Voss, R., Schlatt, S., Grindberg, R. V., & Scholer, H. R. (2017). Single-cell gene expression analysis reveals diversity among human spermatogonia. *Molecular Human Reproduction, 23*, 79–90.

Novershtern, N., Subramanian, A., Lawton, L. N., Mak, R. H., Haining, W. N., McConkey, M. E., Habib, N., Yosef, N., Chang, C. Y., Shay, T., et al. (2011). Densely interconnected transcriptional circuits control cell states in human hematopoiesis. *Cell, 144*, 296–309.

Ohinata, Y., Ohta, H., Shigeta, M., Yamanaka, K., Wakayama, T., & Saitou, M. (2009). A signaling principle for the specification of the germ cell lineage in mice. *Cell, 137*, 571–584.

Okuda, A., Fukushima, A., Nishimoto, M., Orimo, A., Yamagishi, T., Nabeshima, Y., Kuro-o, M., Nabeshima, Y., Boon, K., Keaveney, M., et al. (1998). UTF1, a novel transcriptional coactivator expressed in pluripotent embryonic stem cells and extra-embryonic cells. *The EMBO Journal, 17*, 2019–2032.

Ooi, S. K., & Bestor, T. H. (2008). The colorful history of active DNA demethylation. *Cell, 133*, 1145–1148.

Perrett, R. M., Turnpenny, L., Eckert, J. J., O'Shea, M., Sonne, S. B., Cameron, I. T., Wilson, D. I., Rajpert-De Meyts, E., & Hanley, N. A. (2008). The early human germ cell lineage does not express SOX2 during in vivo development or upon in vitro culture. *Biology of Reproduction, 78*, 852–858.

Robinson, L. L., Gaskell, T. L., Saunders, P. T., & Anderson, R. A. (2001). Germ cell specific expression of c-kit in the human fetal gonad. *Molecular Human Reproduction, 7*, 845–852.

Saliba, A. E., Westermann, A. J., Gorski, S. A., & Vogel, J. (2014). Single-cell RNA-seq: Advances and future challenges. *Nucleic Acids Research, 42*, 8845–8860.

Shapiro, E., Biezuner, T., & Linnarsson, S. (2013). Single-cell sequencing-based technologies will revolutionize whole-organism science. *Nature Reviews Genetics, 14*, 618–630.

Tanay, A., & Regev, A. (2017). Scaling single-cell genomics from phenomenology to mechanism. *Nature, 541*, 331–338.

Tang, F., Lao, K., & Surani, M. A. (2011). Development and applications of single-cell transcriptome analysis. *Nature Methods, 8*, S6–11.

Valli, H., Sukhwani, M., Dovey, S. L., Peters, K. A., Donohue, J., Castro, C. A., Chu, T., Marshall, G. R., & Orwig, K. E. (2014). Fluorescence- and magnetic-activated cell sorting strategies to isolate and enrich human spermatogonial stem cells. *Fertility and Sterility, 102*(566–580), e567.

Virant-Klun, I., Knez, K., Tomazevic, T., & Skutella, T. (2013). Gene expression profiling of human oocytes developed and matured in vivo or in vitro. *BioMed Research International, 2013*, 879489.

Virant-Klun, I., Leicht, S., Hughes, C., & Krijgsveld, J. (2016). Identification of maturation-specific proteins by single-cell proteomics of human oocytes. *Molecular & Cellular Proteomics: MCP, 15*, 2616–2627.

von Kopylow, K., Kirchhoff, C., Jezek, D., Schulze, W., Feig, C., Primig, M., Steinkraus, V., & Spiess, A. N. (2010). Screening for biomarkers of spermatogonia within the human testis: A whole genome approach. *Human Reproduction, 25*, 1104–1112.

von Kopylow, K., Staege, H., Schulze, W., Will, H., & Kirchhoff, C. (2012a). Fibroblast growth factor receptor 3 is highly expressed in rarely dividing human type a spermatogonia. *Histochemistry and Cell Biology, 138*, 759–772.

von Kopylow, K., Staege, H., Spiess, A. N., Schulze, W., Will, H., Primig, M., & Kirchhoff, C. (2012b). Differential marker protein expression specifies rarefaction zone-containing human Adark spermatogonia. *Reproduction, 143*, 45–57.

von Kopylow, K., Schulze, W., Salzbrunn, A., & Spiess, A. N. (2016). Isolation and gene expression analysis of single potential human spermatogonial stem cells. *Molecular Human Reproduction, 22*, 229–239.

Wagner, A., Regev, A., & Yosef, N. (2016). Revealing the vectors of cellular identity with single-cell genomics. *Nature Biotechnology, 34*, 1145–1160.

Woodworth, M. B., Girskis, K. M., & Walsh, C. A. (2017). Building a lineage from single cells: Genetic techniques for cell lineage tracking. *Nature Reviews Genetics, 18*, 230–244.

Yosef, N., & Regev, A. (2011). Impulse control: Temporal dynamics in gene transcription. *Cell, 144*, 886–896.

Advs Exp. Medicine, Biology - Innovations in Cancer Research and Regenerative Medicine (2018) 1: 89–102
DOI 10.1007/5584_2017_116

Published online: 15 November 2017

Cancer Stem Cells in Head and Neck Carcinomas: Identification and Possible Therapeutic Implications

Elize Wolmarans, Sonja C. Boy, Sulette Nel, Anne E. Mercier, and Michael Sean Pepper

Abstract

The recurrence and/or lack of response of certain tumors to radio- and chemotherapy has been attributed to a small subpopulation of cells termed cancer stem cells (CSCs). CSCs have been identified in many tumors (including solid and hematological tumors). CSCs are characterized by their capacity for self-renewal, their ability to introduce heterogeneity within a tumor mass and its metastases, genomic instability, and their insensitivity to both radiation and chemotherapy. The latter highlights the clinical importance of studying this subpopulation since their resistance to traditional treatments may lead to metastatic disease and/or tumor relapse. Head and neck squamous cell carcinomas (HNSCCs) are the sixth most common malignancy worldwide with the highest incidence occurring in East Asia and eastern and southern Africa. Several cellular subpopulations believed to have CSC properties have been isolated from HNSCCs, but at present, identification and characterization of CSCs remains an experimental challenge with no established or standardized protocols in place to confirm their identity. In this review we discuss current approaches to the study of CSCs with a focus on HNSCCs, particularly in the context of what this might mean from a therapeutic perspective.

E. Wolmarans and M.S. Pepper (✉)
Institute for Cellular and Molecular Medicine (ICMM), Department of Immunology, and SAMRC Extramural Unit for Stem Cell Research and Therapy, Faculty of Health Sciences, University of Pretoria, Pretoria, South Africa
e-mail: elizewol@yahoo.com; michael.pepper@up.ac.za

S.C. Boy
Department of Oral Pathology, School of Oral Health Sciences, Sefako Makgatho Health Sciences University, Ga-Rankuwa, Pretoria, South Africa
e-mail: Sonja.iaop@gmail.com

S. Nel
Department of Oral Pathology and Oral Biology, School of Dentistry, Faculty of Health Sciences, University of Pretoria, Pretoria, South Africa
e-mail: sulette.nel@up.ac.za

A.E. Mercier
Department of Physiology, Faculty of Health Sciences, University of Pretoria, Pretoria, South Africa
e-mail: joji.mercier@up.ac.za

Keywords
Cancer stem cells • Head and neck carcinomas

Abbreviations

ABC	ATP-binding cassette
AKT	Protein kinase B
ALCAM	Activated leukocyte cell adhesion molecule
ALDH	Aldehyde dehydrogenase
ATP	Adenosine triphosphate
BCRP	Breast cancer resistant protein
BMI1	Moloney murine leukemia virus insertion site 1
CD44	Cluster of differentiation
CSC	Cancer stem cell
EMT	Epithelial-mesenchymal transition
ESCC	Esophageal squamous cell carcinoma
FACS	Fluorescence-activated cell sorting
HIF	Hypoxia-inducible factors
HNC	Head and neck carcinoma
HNSCC	Head and neck squamous cell carcinoma
HPV	Human papillomavirus
HSA	Heat stable antigen
ICAM1	Intercellular adhesion molecule 1
MAPK	Mitogen-activated protein kinases
NOD	Nonobese diabetic
Oct3/4	Octamer-binding transcription factor 3/4
OSCC	Oral squamous cell carcinoma
P-gp	P-glycoprotein
PI3K	Phosphatidylinositol-3-kinase
POU	Pit-Oct-Unc
SCC	Squamous cell carcinoma
SCID	Severe combined immunodeficiency
SOX2	Sex-determining region Y-box2
SP	Side population

1 Introduction

Head and neck carcinomas (HNCs) comprise a group of malignant epithelial tumors originating at various anatomical sites with a variety of both known and uncertain etiologies. Anatomical areas affected include the oral cavity, oropharynx, pharynx, upper respiratory track, and larynx. Of all malignancies that affect the head and neck, about 90% are histological variants of squamous cell carcinoma (SCC) (Jemal et al. 2011). Based on 2012 global cancer statistics, head and neck squamous cell carcinomas (HNSCCs) are classified as the sixth most common malignancy worldwide with the highest incidence occurring in East Asia and eastern and southern Africa (Torre et al. 2015).

The majority of HNSCCs are strongly associated with environmental and lifestyle risk factors such as obesity, poor nutrition, high tobacco usage, and alcohol consumption (Bruni et al. 2017; Torre et al. 2015; Vigneswaran and Williams 2014). However, an increasing incidence of HNSCCs at specific sites suggests that other etiological factors may be involved. Human papillomavirus (HPV) infection is now a well-established cause of cervical cancer, and there is growing evidence that HPV is an important causative factor in HNSCCs (Bruni et al. 2017; Torre et al. 2015). The pathogenic role of HPV in subsets of head and neck tumors has been recognized by the World Health Organization through reclassification of SCC of the oropharynx (base of the tongue, tonsils, and adenoids) into HPV-positive and HPV-negative SCC (El-Naggar et al. 2017).

Despite ongoing advances in treatment regimes, the prognosis for HNSCCs has improved only marginally over the past 30 year, with the 5-year survival rate ranging from 10% to 50% depending on a number of factors including the clinical stage of the tumor at the time of diagnosis (Karimnejad et al. 2016; Yan et al. 2013). Standard therapy for HNSCCs includes a combination of surgery, chemotherapy, and radiotherapy, which is determined by the histological variety, grade, and clinical staging of the tumor (Zhu et al. 2015). However, irrespective of the treatment regime, patients exhibit relatively high rates of local recurrence as well as regional cervical lymph node and distant tumor metastases, which contribute to significant morbidity and mortality (Karimnejad et al. 2016; Zhu et al. 2015). Recurrent tumor and metastatic deposits are frequently resistant to the original treatment modalities utilized (Karimnejad et al. 2016; Zhu et al. 2015).

The recurrence and lack of response to further radio- and chemotherapy in these tumors has been attributed to a small subpopulation of tumor-initiating cells known as cancer stem cells (CSCs). This subpopulation is believed to (a) have unlimited capacity for self-renewal, (b) introduce heterogeneity into tumors as a result of genomic instability, and (c) show resistance to traditional anticancer treatments with the exception of surgery (Fábián et al. 2013; Gil et al. 2008; Qian et al. 2016).

Traditional approaches to treating HNSCCs are based on a stochastic model of cancer. According to this model, HNSCCs result from the accumulation of genetic changes via successive mutations in different cells and gradual selection of malignant clones (Gil et al. 2008). Cells in the dominant clonal population (s) possess similar tumorigenic potential (Karimnejad et al. 2016), and current therapies mainly target these cells which form the bulk of the tumor.

The CSC model suggests a unidirectional hierarchical organization of cells starting with a progenitor cell endowed with stem cell-like properties (the CSC) (Valent et al. 2012; Visvader and Lindeman 2012; Allegra and Trapasso 2012). The CSC model accommodates the heterogeneity seen in human tumors and postulates that this heterogeneity is due to distinctive characteristics of different CSCs present in the primary tumor. CSCs share similar features to normal stem cells, including the ability to selectively self-renew, proliferate, and differentiate into the cell mass that compromises the bulk of the tumor (Valent et al. 2012; Visvader and Lindeman 2012; Allegra and Trapasso 2012). More importantly, it is believed that their stem cell-like features provide CSCs with a measure of protection against traditional anticancer therapies compared to the rest of the cells in the tumor (Fábián et al. 2013), thereby providing one possible explanation for treatment resistance and/or tumor relapse. This highlights the clinical importance of identifying and defining these subpopulations of cells.

Much effort has been invested into characterizing and identifying CSCs in various tumor types and exploring the mechanisms of their involvement in the unique reaction of certain tumors to different treatment regimes. At present, characterization and identification of CSCs is largely limited to the experimental setting, in which CSCs demonstrate (retrospectively) the capability of generating and propagating a malignant cell population. In this review we discuss current approaches to the identification and study of CSCs in HNSCCs, particularly from a therapeutic perspective.

2 Identification of CSCs

Three main strategies are used to isolate putative CSCs from solid tumors: (i) tumorsphere formation assays performed under nonattached culture conditions, (ii) side population assays aimed at identifying cells with greater efflux potential, and (iii) fluorescence-activated cell sorting (FACS) where subpopulations are sorted based on known cell surface markers. The subpopulations identified and isolated are then subjected to further testing that assesses their tumorigenic ability and self-renewal properties in vitro and/or in vivo.

2.1 Sphere Formation Assays for the Identification of Putative CSCs Populations in HNSCC

For solid tumors, the most accessible method for studying self-renewal in vitro is tumorsphere formation under nonattached culture conditions (Valent et al. 2012; Zhang et al. 2012a; Fábián et al. 2013). Primary spheres are generally assessed for their ability to form secondary spheres and their ability to form colonies once plated onto adherent plates (Harper et al. 2007). This method relies on the theory that isolated cancer cells that form tumorspheres are capable of proliferation, self-renewal, and greater tumorigenic capacity (Zhang et al. 2012a). Sphere culturing is thus used to isolate, enrich, and expand putative CSC populations.

The tumorsphere formation assay has been used with both HNSCC cell lines and primary tumor tissue to identify and study various subpopulations (Chiou et al. 2008; Harper et al. 2007; Zscheppang et al. 2016; Yu et al. 2010). Tumorspheres have a higher proliferative and self-renewal capacity than their parent populations in vitro (Chiou et al. 2008; Harper et al. 2007). They also have distinctive phenotypes compared to the parent populations, showing high positivity for markers such as cluster of differentiation (CD) 44, CD133, aldehyde dehydrogenase 1 (ALDH1), and breast cancer resistant protein (BCRP/ABCG2) (Chiou et al. 2008; Harper et al. 2007; Yu et al. 2010; Chen et al. 2009). When compared to the parent cell population, these spheres have also been shown to have greater survival abilities against ionizing radiation in vitro (Chiou et al. 2008; Zscheppang et al. 2016) and greater tumorigenic capacity following xenotransplantation (Chiou et al. 2008; Zscheppang et al. 2016).

Advantages of this method include lower cost and quicker results but several shortcomings have been identified. The in vitro system is unable to replicate the three-dimensional structure and environment of a tumor in vivo. Factors important for in vivo growth and self-renewal of some or all CSCs may not be provided by the in vitro conditions. Additionally, the growth of the cells in vitro is usually limited to several weeks or months, during which time cells may differentiate, die, or transform (Valent et al. 2012).

2.2 Side Population Discrimination Assays for the Identification of CSC in HNSCC

Another strategy used to identify putative CSC populations is based on the ability of these cells to efflux a fluorescent dye that can bind to DNA. One possible mechanism whereby CSCs remain resistant to traditional anticancer chemotherapeutics lies in their cell surface expression of adenosine triphosphate (ATP)-binding cassette (ABC) transporter proteins. These proteins actively translocate or efflux cytotoxic chemicals out of cells (Dean 2009). In the assay, the cells are stained with a fluorescent dye such as Hoechst 33,342, followed by incubation for a period of time. When analyzed via flow cytometry, the subpopulation of cells with efflux capability, believed to be CSCs, will have lower fluorescence than the more differentiated tumor and stromal cells. This subpopulation is termed the side population (SP).

The SP assay was first described by Goodell et al. when it was used to isolate a subpopulation highly enriched in hematopoietic stem cells (Goodell et al. 1996). Since then, the SP assay has been used in many human tissues in an attempt to identify somatic stem cells, as well as in a variety of cancers in an attempt to identify a subpopulation of cells with CSC characteristics.

The SP assay has also been used to identify putative CSC populations in HNSCCs using primary tumor cell isolates as well as established cell lines. Zhang et al. tested for the presence of the SP in nine oral squamous cell carcinoma (OSCC) cell lines including Tca8113, NTRC, and TSCC and 11 primary OSCC specimens. They found the SP to be present in all cell lines and primary tumors (Zhang et al. 2009). Various other studies have further shown that SP cells are resistant to chemotherapeutic drugs and have greater colony formation abilities than non-SP cells in vitro, as well as greater clonogenicity and tumorigenicity following xenotransplantation (Zhang et al. 2009, 2012a; Li et al. 2011).

There are, however, limitations to the SP assay. It is a highly technical assay which requires modification for each cell type studied (Golebiewska et al. 2011). Results show poor reproducibility due to high interindividual variability between samples and laboratory protocols (Golebiewska et al. 2011). Even though Zang et al. found the SP in all OSCC cell lines and primary tumors tested, the authors reported variable percentages of SP cells within the samples ranging from 0.1% to 10% (Zhang et al. 2009). It is also important to emphasize that the ABC transporters and the SP phenotype are not exclusive to stem cells (Golebiewska et al. 2011). Since all tumors are heterogeneous in nature, there is no guarantee that the SP only represents CSCs and may in all likelihood contain both CSCs and normal stromal cells with efflux capabilities. The SP assay, similar to the sphere formation assay, may however be useful as an enrichment step in order to obtain a more homogeneous subpopulation to investigate for stemlike properties.

2.3 Identification of CSC in HNSCCs Using Specific Markers

The third method of CSC isolation consists of sorting cells based on either phenotypic and/or transcriptional profiles. However, a consensus on unequivocal and specific markers for the identification of CSCs in all cancers types has not yet been achieved (Visvader and Lindeman 2012; Allegra and Trapasso 2012; Yan et al. 2013).

An extensive list of possible CSC markers has been studied in HNSCCs. Below we discuss data from the more promising markers described in the literature. Table 1 provides a summary of the markers discussed.

2.3.1 ABC Transporters

ABC transporters form one of the largest families of membrane transporter proteins (Falasca and Linton 2012; Huls et al. 2009; Lin et al. 2006). The human ABC transporter protein family consists of 49 members, which are divided into seven subfamilies (A to G) based on similarities in gene structure, order of domains, and sequence homology (Huls et al. 2009; Falasca and Linton 2012; Dean et al. 2001; Leonard et al. 2003). This diverse family plays an important role in the maintenance of cellular homeostasis through the transport of lipids and organic anions, iron metabolism, and providing tissue protection via drug resistance and metabolism (Huls et al. 2009; Erdei et al. 2014; Leonard et al. 2003; Lin et al. 2006). Expression of the transporters is tightly regulated, and their classification is based on their ability to recognize various toxic agents. They are able to functionally substitute for each other, thereby emphasizing the importance of their protective role (Huls et al. 2009; Erdei et al. 2014).

BCRP/ABCG2 is expressed by many tumors and has been linked to tumor initiation and promotion as well as tumor cell proliferation (Qian et al. 2016; Ding et al. 2010). It is also an important multidrug resistance protein shown to confer cross-resistance on several classes of cancer chemotherapeutic agents (Mao and Unadkat 2015). Because of this function, ABCG2 expression has been linked to the SP assay and is believed to be one of the main mediators of the SP phenotype along with other members of this family such as P-glycoprotein (P-gp/ABCB1).

BCRP/ABCG2 expression has been demonstrated in putative HNSCC CSC subpopulations (Hang et al. 2012; Li et al. 2011; Tsai et al. 2011) along with other members of the family such as ABCA3 (Li et al. 2011), P-gp/ABCB1 (Zhang et al. 2009, 2012a; Li et al. 2011), ABCC1 (Li et al. 2011), and ABCB5 (Grimm et al. 2012).

The expression of these proteins has also been associated with poor patient prognosis. High levels of expression of both ABCG2 and ABCB5 are associated with tumor progression and recurrence (Grimm et al. 2012; Hang et al. 2012).

2.3.2 CD44

CD44 is the most well-established and frequently used CSC marker and has been

Table 1 Markers studied with the aim of identifying CSC subpopulations in HNSCCs

Marker name / *Marker abbreviation*	Expression in HNSCCs	Prognostic value	Selected references
ABC transporters			
ABC subfamily A member 3 / *ABCA3*	Yes	Unknown	Li et al. (2011), Albrecht and Viturro (2007) and Vasiliou et al. (2009)
P-glycoprotein / *P-gp/ABCB1*	Yes	Unknown	Huls et al. (2009), Lin et al. (2006), Wang et al. (2000) and Zhang et al. (2009, 2012a)
ABC subfamily B member 5 / *ABCB5*	Yes	Yes	Grimm et al. (2012)
Multidrug resistance protein 1 / *MRP1/ABCC1*	Yes	Unknown	Li et al. (2011) and Borst et al. (1999)
Breast cancer resistance protein / *BCRP/ABCG2*	Yes	Yes	Lin et al. (2006), Zhang et al. (2009, 2012a), Chiou et al. (2008), Tsai et al. (2011) and Hang et al. (2012)
Extracellular markers			
Hyaluronan receptor / *CD44*	Controversial	Yes	Qian et al. (2016), Prince et al. (2007), Mărgăritescu et al. (2012) and Zhang et al. (2009)
CD24	Yes	Yes	Koukourakis et al. (2012), Satpute et al. (2013), Zimmerer et al. (2017) and Han et al. (2014)
Prominin-1 / *CD133*	Controversial	Controversial	Mărgăritescu et al. (2012), Okamoto et al. (2013), Hang et al. (2012), Tsai et al. (2011) and Chiou et al. (2008)
Intracellular markers			
Aldehyde dehydrogenase / *ALDH1*	Controversial	Controversial	Zhang et al. (2012a), Clay et al. (2010), Koukourakis et al. (2012) and Yata et al. (2015)
Moloney murine leukemia virus insertion site 1 / *BMI1*	Yes	Controversial	Yu et al. (2010), Tsai et al. (2011) and Fan et al. (2017)
Nanog	Yes	Yes	Zhang et al. (2012a, b), Chiou et al. (2008), Tsai et al. (2011) and Fan et al. (2017)
Oct3/4	Yes	Yes	Zhang et al. (2012a, b), Chiou et al. (2008), Tsai et al. (2011) and Fan et al. (2017)
Sex-determining region Y box 2 / *SOX-2*	Yes	Yes	Zhang et al. (2012b) and Fan et al. (2017)

intensively studied in HNSCCs. It is a cell surface transmembrane protein that belongs to the adhesion molecule superfamily (Qian et al. 2016). This protein regulates cell-cell interactions, cell migration, and cell adhesion (Mărgăritescu et al. 2012). It is implicated in tissue development, inflammation, and wound healing and is also believed to influence tumorigenesis and progression, playing a role in invasion and metastasis (Qian et al. 2016). The molecule activates a variety of receptor tyrosine kinases in many cancer types, through which it drives increased proliferation and survival of tumor cells via activation of the mitogen-activated protein kinases (MAPK) and phosphatidylinositol-3-kinase/protein kinase B (PI3K/AKT) pathways (Mărgăritescu et al. 2012; Misra et al. 2006).

Prince et al. were the first to report that HNSCCs contain two distinct populations of either CD44$^+$ or CD44$^-$ cells (Prince et al. 2007). Cells of the CD44$^+$ subpopulation were able to initiate tumor growth in xenografts more efficiently than their CD44$^-$ counterparts (Prince et al. 2007). In addition, tumors developed from the transplantations displayed phenotypically diverse populations of cells, indicating that the CD44$^+$ cells were not only able to initiate tumor growth but were also able to differentiate into phenotypically diverse daughter cells (Prince et al. 2007).

Since the publication of this landmark study, CD44 has been used routinely as a marker in experiments on HNSCC CSC subpopulations (Zhang et al. 2009; Noto et al. 2013; Chen et al. 2009; Harper et al. 2007). CD44 expression has also been correlated with a greater frequency of lymph node metastasis, higher primary tumor recurrence, resistance to radiotherapy, and a poor clinical prognosis in HNSCC patients (Wang et al. 2009; Han et al. 2014).

More recently however, studies have called into question the sensitivity and specificity of CD44 as a CSC surface marker in HNSCCs, as it is expressed on the majority of cells that constitute head and neck tissues, including normal oral epithelium as well as potentially malignant and malignant lesions (Mărgăritescu et al. 2012; Yan et al. 2013).

2.3.3 CD133

CD133 is a transmembrane pentaspan glycoprotein specifically localized on cellular protrusions (Mărgăritescu et al. 2012). The function of this protein is not fully known although it is used as a marker of CSCs in many solid tumors including the brain, colon, pancreas, prostate, liver, lung, and kidney (Qian et al. 2016; Mărgăritescu et al. 2012; Okamoto et al. 2013).

The literature regarding CD133 expression in HNSCCs lacks unanimity. Tsai et al. demonstrated CD133 positivity in a subpopulation of OSCCs which exhibited resistance to the conventional chemotherapeutic drug cisplatin in vitro (Tsai et al. 2011). Zhang et al. reported on a CD133$^+$ subpopulation within OSCC cell lines which exhibits higher clonogenicity and tumorigenicity in xenograft models when compared to their CD133$^-$ counterparts (Zhang et al. 2010).

In contrast to these findings, other investigators failed to demonstrate CD133 expression in freshly isolated OSCC from patient biopsy samples (Mărgăritescu et al. 2012; Okamoto et al. 2013). In a study done by Hang et al., CD133 was found in 27% of patients with esophageal squamous cell carcinoma (ESCC) in which the proportion of CD133$^+$ cells in ESCCs ranged from 3% to 46% with higher percentages being expressed in the more differentiated tumors (Hang et al. 2012).

In addition, no consensus has been reached on the prognostic value of CD133. An association between CD133 expression and poor prognosis has been reported in HNSCC patients (Chiou et al. 2008; Fan et al. 2017). In contrast, Hang et al. found no significant association between CD133 expression and the 5-year survival rate of ESCC patients (Hang et al. 2012).

2.3.4 CD24

CD24, also known as heat stable antigen 24 (HSA), is a mucin adhesion molecule and is expressed by pre-B lymphocytes and neutrophils (Satpute et al. 2013). It has been identified as a ligand of P-selectin, an adhesion receptor found on activated platelets and endothelial cells (Satpute et al. 2013; Modur et al. 2016). This molecule is associated with tumor growth and metastasis (Han et al. 2014) and has been shown to be expressed in HNSCCs (Zimmerer et al. 2017; Koukourakis et al. 2012; Modur et al. 2016).

Studies on HNSCC cell lines and primary tumor cells have revealed the presence of a subpopulation of cells that is CD24$^+$. This subpopulation has been shown to have a greater ability to self-renew and differentiate and demonstrates a greater resistance to chemotherapeutic drugs such as gemcitabine and cisplatin in vitro (Han et al. 2014; Modur et al. 2016). In addition Zimmer et al. showed that CD24$^+$ subpopulations, isolated from freshly isolated OSCC samples, could initiate tumor growth and accelerate angiogenesis in vivo (Zimmerer et al. 2017). The study showed greater functional capillary density of newly formed microvessels by the CD24^{+} subpopulation compared to the CD24^{-} subpopulation after the cells were transplanted into nonobese diabetic/severe combined immunodeficient (NOD/SCID) mice (Zimmerer et al. 2017). The authors believe that CD24^{+} cells have the potential to induce and modulate angiogenesis and to stimulate other tumor cells to switch to an angiogenic phenotype (Zimmerer et al. 2017).

From a prognostic point of view, the expression of CD24 along with CD44 is associated with lower overall survival rates compared to other phenotypes (Han et al. 2014; Modur et al. 2016).

2.3.5 ALDH1

ALDH enzymes are a family of cytosolic isoenzymes that are involved in cell differentiation and detoxification and confer drug resistance via oxidation of intracellular aldehydes (Zhang et al. 2012b; Clay et al. 2010). The ALDH family member most extensively investigated is ALDH1, and it is regarded as a marker for hematopoietic stem and progenitor cells (Qian et al. 2016; Zhang et al. 2012b). ALDH1 has also been identified as a potential functional marker for CSCs as it is expressed in various tumor types (Clay et al. 2010).

ALDH1 expression is significantly increased in subpopulations of ESCC cell lines (Zhang et al. 2012a; Yata et al. 2015) and in primary isolated tumors from HNSCC patients (Yu et al. 2010; Chen et al. 2009).

ALDH1^{+} subpopulations from HNSCC primary cell isolates possess greater tumorsphere- and colony-forming capacity in vitro than their ALDH1^{-} counterparts (Yu et al. 2010; Chen et al. 2009; Yata et al. 2015). In addition, ALDH1^{+} subpopulations are highly tumorigenic when compared to ALDH1^{-} subpopulations following xenotransplantation (Yu et al. 2010; Chen et al. 2009; Clay et al. 2010; Yata et al. 2015).

Although ALDH1 is present in HNSCC subpopulations, there is controversy as to whether ALDH1 expression may be viewed as a prognostic marker. Whereas some studies associate ALDH1 protein expression with lymph node metastasis and poor survival (Wang et al. 2012), others report that ALDH1 expression in HNSCCs is correlated with a favorable clinical outcome (Koukourakis et al. 2012).

2.3.6 BMI1

Moloney murine leukemia virus insertion site 1 (BMI1) is a member of the Polycomb (PcG) family of transcriptional repressors that mediate gene silencing by mono-ubiquitination of histone H2A (Yu et al. 2010; Chen et al. 2017). This gene plays a role in the regulation of the cell cycle and the regeneration of hematopoietic and neuronal stem cells (Yu et al. 2010).

BMI1 has been implicated in tumorigenesis in several human cancer types including HNSCC (Prince et al. 2007; Yu et al. 2010). Yu et al. studied BMI1 expression in subpopulations of HNSCCs and found that BMI1^{+} subpopulations have enhanced tumorigenic abilities both in vitro and in vivo (Yu et al. 2010). When BMI1 was knocked down, there was a decrease in tumorigenicity in vitro, while inducing overexpression of BMI1 resulted in increased tumor growth, metastatic activity, and radioresistance in vivo (Yu et al. 2010). In another study, subpopulations of OSCCs with increased BMI1 expression exhibited resistance to the chemotherapeutic drug cisplatin in vitro (Tsai et al. 2011).

Studies evaluating BMI1 as a prognostic marker for HNSCCs have shown inconsistent results. Yu et al. reported that BMI1 expression correlates with poor overall patient survival (Yu et al. 2010). However, a recent meta-analysis could not identify an association between the expression of BMI1 in HNSCC patients and overall survival (Fan et al. 2017).

2.3.7 Oct3/4, SOX2, and Nanog

The transcription factors octamer-binding transcription factor 3/4 (Oct3/4), sex-determining region Y-box2 (SOX2), and Nanog play essential roles in the maintenance of pluripotency and self-renewal (Zhang et al. 2012b; González-Moles et al. 2013).

Oct3/4 belongs to the Pit-Oct-Unc (POU) transcription factor family (Bourguignon et al. 2012). It is the product of the OTF3 gene and has been shown to maintain stemness in pluripotent stem cells (Eun et al. 2017). Oct3/4 functions as a master regulator during embryonic development by interacting with other embryonic regulators such as SOX2 and Nanog to oversee a vast regulatory network that maintains pluripotency and inhibits differentiation (Bourguignon et al. 2012). It is considered to be one of the best indicators of stemness (González-Moles et al. 2013).

SOX2 belongs to the family of high-mobility group transcription factors that play important roles in stem cell function, development, and organogenesis (Bourguignon et al. 2012; Ren et al. 2016).

Nanog is another important transcription factor involved in self-renewal and the maintenance of pluripotency (Bourguignon et al. 2012). Nanog signaling is regulated by interactions between pluripotent stem cell regulators (such as SOX2 and Oct3/4) which together control the expression of a set of target genes required for pluripotency (Bourguignon et al. 2012).

These factors are known to form a self-organized core of transcription factors that maintain pluripotency and self-renewal (Bourguignon et al. 2012). All three markers have been studied in HNSCCs, and all three have been found to be highly expressed in putative CSC populations (Tsai et al. 2011; Bourguignon et al. 2012; Kaseb et al. 2016). Tsai et al. demonstrated that subpopulations of OSCC cells isolated via tumorsphere formation expressed elevated levels of Oct4 and Nanog and displayed increased resistance to the chemotherapeutic drug cisplatin in vitro (Tsai et al. 2011). Similar results were obtained by another research group who found Oct3/4, SOX2, and Nanog gene expression to be significantly higher in putative CSC subpopulations (Bourguignon et al. 2012).

Elevated expression of these markers in HNSCC has also been associated with a poor prognosis and lower overall survival rate (Chiou et al. 2008; Fan et al. 2017; Koukourakis et al. 2012).

2.4 Xenografts

Xenografts of candidate CSC populations into immunodeficient mice remain the gold standard for verification of stem cell properties by assessing the tumor initiation capabilities of the specific isolated subpopulation (Valent et al. 2012; Visvader and Lindeman 2012; Zhang et al. 2012b). This method has been successfully applied to the detection of CSCs in many types of human malignancies including HNSCCs (Prince et al. 2007; Wang et al. 2007; Chiou et al. 2008; Chen et al. 2009; Zhang et al. 2009).

However, the ability or failure of potential CSCs to produce a detectable malignant population in a transplanted mouse may not accurately reflect the behavior of the cells within the patient's original tumor micromilieu, and thus caution should be used when interpreting results (Valent et al. 2012; Zhang et al. 2012b). As the majority of xenotransplantation studies utilize immunodeficient mice, limitations of these models must be expected. These include lack of cytokines which can stimulate the growth of CSCs, the inability to precisely model the tumor-specific microenvironment, and the deficiency of natural immunosurveillance (Valent et al. 2012). Humanized mice, with a more human-type hematopoietic and immune microenvironment, may have to be considered for future studies.

3 Clinical Challenges of CSC in the Treatment of HNSCCs

The small subpopulation of CSCs within the diverse and heterogeneous cell population comprising a tumor mass may be responsible for tumor recurrence, the promotion of metastasis due to high migration capacity, as well as resistance to both cytotoxic agents and radiation therapy. Intrinsic characteristics such as an increase in ABC transmembrane proteins, a semi-quiescent state, and altered apoptotic mechanisms within CSCs limit susceptibility to cell death (Clarke and Fuller 2006; Zhang et al. 2009).

Therapeutic success following radiation of solid tumors is inversely proportional to the percentage of CSCs within the tumor mass (Yaromina et al. 2007). CSCs not eliminated by radiation are potentially responsible for non-remission and recurrence, as they have the capacity to renew and differentiate into the heterogeneous constituents of the tumor (Wicha et al. 2006). CSCs themselves are inherently more radioresistant, employing mechanisms which may encompass increased checkpoint activation and enhanced DNA damage repair responses (Eyler and Rich 2008). Increasing the radiation dose in HNSCCs is however associated with intolerable side effects such as xerostomia (Toledano et al. 2012). This may be explained by the radiation affecting microniches in which normal salivary stem cells are located, often in close proximity to blood vessels concentrated in the salivary glands (Vissink et al. 2010). Mechanisms sparing normal tissues aimed at curtailing the dose-limiting side effects of radiation are of interest to many researchers (Vallard et al. 2016).

Central tumor hypoxia seen in larger masses may also confer a survival advantage to CSCs being treated with chemotherapeutics or radiation (Heddleston et al. 2010). In methods similar to the hypoxic maintenance of the pluripotency of embryonic stem cells, poor perfusion of larger tumor masses may promote the CSC phenotype by creating specific CSC niches. Suboptimal blood flow not only limits the distribution of chemotherapeutic agents to cancer cells but also decreases the oxygen tension required for free radical formation in response to radiation (Satpute et al. 2013). The overexpression of hypoxia-inducible factors (HIFs) in CSCs may also contribute to radioresistance of HNSCCs (Vlashi et al. 2009). Overexpression of HIF-1-α in CSCs mediates the induction of epithelial-mesenchymal transitions (EMT) in complex interactions which confer increased mobility on CSCs, as well as ensuring the maintenance of their pluripotency (Yang et al. 2010). Interactions between stem cells and stromal components in EMT which enhance local invasion and metastasis appear to involve the Wnt/beta-catenin signaling pathway (Sato et al. 2004; Zechner et al. 2003; Takahashi-Yanaga and Kahn 2010).

Treatment of HNSCCs still mainly involves surgical removal of the primary lesion wherever possible, with particular care being taken to include liberal surgical margins together with case-specific permutations in perioperative chemo- and/or radiation therapy (Satpute et al. 2013). The latter two modalities also aim at targeting residual tumor tissue. These debulking strategies however do not ensure elimination of CSCs, and these treatments may therefore fail to ensure remission and may predispose the patient to disease recurrence, as is often seen in HNSCCs. Future strategies should therefore aim to target these CSCs or the CSC niche specifically, which, when done in combination with traditional treatment modalities, might ensure greater long-term disease remission. Mechanisms which promote sensitivity to cytotoxic agents and radiation, prevent evasion of immune surveillance, and/or inhibit EMT crosstalk pathways may also hold the key to improved treatment outcomes in HNSCCs (Méry et al. 2016). As the mechanisms targeted are often integral to normal tissue stem cell function, treatment may result in severely toxic side effects.

4 Comments and Future Perspectives

The main challenge remains the identification of CSC subpopulations within primary tumors. The heterogeneity of these tumors is well known (Liang and Fu 2017). Techniques such as sphere formation and the side population discrimination assay can only be used to enrich for a population that may contain both cancer cells and CSCs.

Using markers to identify putative CSC populations also has its limitations. The existence of various isoforms of surface markers used for the identification of CSCs should be considered. Different antibody epitopes as well as possible cross-reactions between antibodies need to be considered when designing CSC studies to prevent false positive and/or false negative results (Mărgăritescu et al. 2012). CD44, for example, has various isoforms (Spiegelberg et al. 2014), three of which (CD44-v3, v6 and v10) have been associated with increased metastasis and aggressive HNSCC progression (Wang et al. 2009). The use of different marker isoforms by different research teams may provide one explanation for the contrasting results reported. Another difficulty with regard to CSC surface markers is that many are not only expressed on CSCs but are also present on other cell types including stem cells, somatic cells, and other tumor cells. The expression of these markers may also vary greatly between patients and different tumors.

It is clear that the use of a single marker is currently not sufficient to isolate a pure CSC subpopulation from a given tumor. Refined methods for the identification and characterization of CSCs are therefore needed. Multiple integrated analysis approaches such as transcriptomics and proteomics are being used to study these cells.

Proteomics is a powerful tool for identifying signaling complexes which control CSC differentiation and regulate CSC maintenance pathways (Tsai et al. 2015; Yan et al. 2013). Using mass spectrometry and liquid chromatography-mass spectrophotometry, additional markers such as activated leukocyte cell adhesion molecule (ALCAM/CD166) and intercellular adhesion molecule 1 (ICAM1) have been identified in putative CSC populations of HNCs (Yan et al. 2013; Tsai et al. 2015). It should be noted however that the abovementioned studies and techniques were all conducted on cancer cell lines and not on primary isolated cells from tumor biopsies.

Next-generation sequencing and bioinformatics are also generating a better understanding of the cancer genome. The transcriptome includes signaling and regulatory molecules as well as essential housekeeping molecules and in its totality reflects a cell's identity (Saliba et al. 2014). To date, most transcriptome studies are conducted at a "population level," averaging the transcriptomes of millions of cells. It is now possible to study cells at the single-cell level providing an opportunity to dissect the complexity and heterogeneity of the tumor mass in detail. Increased understanding of tumor heterogeneity now offers researchers the potential to identify and study various subclones within a tumor mass (Jamal-Hanjani et al. 2015). Much work remains to be done to fully elucidate the biology of CSCs in HNSCC. The development of more focused treatment strategies will require a detailed understanding of the biological processes that generate these cells which appear to initiate and drive recurrence and aggressive metastatic tumor spread. Single-cell transcriptome analysis will allow for a more detailed and accurate investigation of CSCs and possible subpopulations; to the authors' knowledge, data generated using this approach in the study of CSCs in HNSCCs has not yet been published.

In conclusion, the study of CSCs in HNSCC is important due to the high rates of recurrence and metastases encountered in these patients. As a result of the complexity and heterogeneity of HNSCCs, the application of techniques such as proteomics and transcriptomics to the study of CSC subpopulations is likely to result in a more efficient therapeutic option for these patients in the long term.

Acknowledgements This research was funded by the South African Medical Research Council in terms of the SAMRC's Flagship Award Project SAMRC-RFA-UFSP-01-2013/STEM CELLS, the SAMRC Extramural Unit for Stem Cell Research and Therapy and the Institute for Cellular and Molecular Medicine of the University of Pretoria.

Conflicts of Interest The authors have no conflicts of interest to declare.

Author Contribution MSP conceived the project, EW drafted the first version of the manuscript, and EW, SB, SN, AEM, and MSP provided intellectual input and contributed to the writing of the manuscript. All authors vetted and approved the final version of the manuscript.

References

Albrecht, C., & Viturro, E. (2007). The ABCA subfamily--gene and protein structures, functions and associated hereditary diseases. *Pflügers Archiv, 453*(5), 581–589.

Allegra, E., & Trapasso, S. (2012). Cancer stem cells in head and neck cancer. *Oncotargets and Therapy, 5*, 375–583.

Borst, P., Evers, R., Kool, M., et al. (1999). The multidrug resistance protein family. *Biochimica et Biophysica Acta, 1461*, 347–357.

Bourguignon, L. Y., Wong, G., Earie, C., et al. (2012). Hyaluronan-CD44v3 interaction with Oct4-Sox2-Nanog promotes miR-302 expression leading to self-renewal, clonal formation, and cisplatin resistance in cancer stem cells from head and neck squamous cell carcinoma. *The Journal of Biological Chemistry, 287*(39), 32800–32824.

Bruni, L., Barrionuevo-Rosas, L., Albero, G. et al. (2017). *Human papillomavirus and related diseases in South Africa*. Summary Report.

Chen, Y. C., Chen, Y. W., Hsu, H. S., et al. (2009). Aldehyde dehydrogenase 1 is a putative marker for cancer stem cells in head and neck squamous cancer. *Biochemical and Biophysical Research Communications, 385*(3), 307–313.

Chen, D., Wu, M., Li, Y., et al. (2017). Targeting BMI1 + cancer stem cells overcomes Chemoresistance and inhibits metastases in squamous cell carcinoma. *Cell Stem Cell, 20*(5), 621–634.

Chiou, S. H., CC, Y., Huang, C. Y., et al. (2008). Positive correlations of Oct-4 and Nanog in oral cancer stem-like cells and high-grade oral squamous cell carcinoma. *Clinical Cancer Research, 14*(13), 4085–4095.

Clarke, M., & Fuller, M. (2006). Stem cells and cancer: Two faces of eve. *Cell, 124*(6), 111–1115.

Clay, M. R., Tabor, M., Owen, J. H., et al. (2010). Single-marker identification of head and neck squamous cell carcinoma cancer stem cells with aldehyde dehydrogenase. *Head & Neck, 23*(9), 1195–1201.

Dean, M. (2009). ABC transporters, drug resistance, and cancer stem cells. *Journal of Mammary Gland Biology and Neoplasia, 14*(1), 3–9.

Dean, M., Hamon, Y., & Chimini, G. (2001). The human ATP-binding cassette (ABC) transporter superfamily. *Journal of Lipid Research, 42*(7), 1007–1017.

Ding, X., Wu, J., & Jiang, C. (2010). ABCG2: A potential marker of stem cells and novel target in stem cell and cancer therapy. *Life Sciences, 86*(17–18), 631–637.

El-Naggar, A., Chan, J., & Grandis, J. (Eds). (2017). Tumours of the oropharynx (base of tongue, tonsils, adenoids). In *WHO Classification of Head and Nect Tumours*. IARC.

Erdei, Z., Lőrincz, R., Szebényi, K., et al. (2014). Expression pattern of the human ABC transporters in pluripotent embryonic stem cells and in their derivatives. *Cytometry Part B, Clinical Cytometry, 86*(5), 299–310.

Eun, K., Ham, S. W., & Kim, H. (2017). Cancer stem cell heterogeneity: Origin and new perspectives on CSC targeting. *BMB Reports, 50*(3), 117–125.

Eyler, C., & Rich, J. (2008). Survival of the fittest: Cancer stem cells in therapeutic resistance and angiogenesis. *Journal of Clinical Oncology, 26*(17), 2839–2845.

Fábián, Á., Vereb, G., & Szöllosi, J. (2013). The hitchhikers guide to cancer stem cell theory: Markers, pathways and therapy. *Cytometry. Part A, 83*(1), 62–71.

Falasca, M., & Linton, K. J. (2012). Investigational ABC transporter inhibitors. *Expert Opinion on Investigational Drugs, 21*(5), 657–666.

Fan, Z., Li, M., Chen, X., et al. (2017). Prognostic value of cancer stem cell markers in head and neck squamous cell carcinoma: A meta-analysis. *Scientific Reports, 7*, 1–8.

Gil, J., Stembalska, A., Pesz, K. A., et al. (2008). Cancer stem cells: The theory and perspectives in cancer therapy. *Journal of Applied Genetics, 49*(2), 193–199.

Golebiewska, A., Brons, N. H., Bjerkvig, R., et al. (2011). Critical appraisal of the side population assay in stem cell and cancer stem cell research. *Cell Stem Cell, 8*(2), 136–147.

González-Moles, M. A., Scully, C., Ruiz-Ávila, I., et al. (2013). The cancer stem cell hypothesis applied to oral carcinoma. *Oral Oncology, 49*(8), 738–746.

Goodell, M. A., Brose, K., Paradis, G., et al. (1996). Isolation and functional properties of murine hematopoietic stem cells that are replicating in vivo. *The Journal of Experimental Medicine, 183*, 1797–1806.

Grimm, M., Krimmel, M., Polligkeit, J., et al. (2012). ABCB5 expression and cancer stem cell hypothesis in oral squamous cell carcinoma. *European Journal of Cancer, 48*(17), 3186–3197.

Han, J., Fujisawa, T., Husain, S. R., et al. (2014). Identification and characterization of cancer stem cells in human head and neck squamous cell carcinoma. *BMC Cancer, 14*(1), 173–184.

Hang, D., Dong, H. C., Ning, T., et al. (2012). Prognostic value of the stem cell markers CD133 and ABCG2 expression in esophageal squamous cell carcinoma. *Diseases of the Esophagus, 25*(7), 638–644.

Harper, L. J., Piper, K., Common, J., et al. (2007). Stem cell patterns in cell lines derived from head and neck squamous cell carcinoma. *Journal of Oral Pathology & Medicine, 36*(10), 594–603.

Heddleston, J. M., Li, Z., Lathia, J. D., et al. (2010). Hypoxia inducible factors in cancer stem cells. *British Journal of Cancer, 102*(5), 789–795.

Huls, M., Russel, F. G., & Masereeuw, R. (2009). The role of ATP binding cassette transporters in tissue defense and organ regeneration. *The Journal of Pharmacology and Experimental Therapeutics, 328*(1), 3–9.

Jamal-Hanjani, M., Quezada, S. A., Larkin, J., et al. (2015). Translational implications of tumor heterogeneity. *Clinical Cancer Research, 21*(6), 1258–1266.

Jemal, A., Bray, F., Center, M. M., et al. (2011). Global cancer statistics. *CA: a Cancer Journal for Clinicians, 61*(2), 69–90.

Karimnejad, K., Lindquist, N., & Lin, R. (2016). The role of cancer stem cells in head and neck squamous cell carcinoma and its clinical implications. In *New aspects in molecular and cellular mechanisms of human carcinogenesis* (pp. 97–113).

Kaseb, H. O., Fohrer-Ting, H., Lewis, D. W., et al. (2016). Identification, expansion and characterization of cancer cells with stem cell properties from head and neck squamous cell carcinomas. *Experimental Cell Research, 348*(1), 75–86.

Koukourakis, M. I., Giatromanolaki, A., Tsakmaki, V., et al. (2012). Cancer stem cell phenotype relates to radio-chemotherapy outcome in locally advanced squamous cell head–neck cancer. *British Journal of Cancer, 106*(5), 846–853.

Leonard, G. D., Fojo, T., & Bates, S. E. (2003). The role of ABC transporters in clinical practice. *The Oncologist, 8*(5), 411–424.

Li, H., Gao, Q., Guo, L., et al. (2011). The PTEN/PI3K/Akt pathway regulates stem-like cells in primary esophageal carcinoma cells. *Cancer Biology & Therapy, 11*(11), 950–958.

Liang, S. B., & Fu, L. W. (2017). Application of single-cell technology in cancer research. *Biotechnology Advances, 35*(4), 443–449.

Lin, T., Islam, O., & Heese, K. (2006). ABC transporters, neural stem cells and neurogenesis – a different perspective. *Cell Research, 16*, 857–871.

Mao, Q., & Unadkat, J. D. (2015). Role of the Breast Cancer Resistance Protein (BCRP/ABCG2) in drug transport—An update. *The AAPS Journal, 17*(1), 65–82.

Mărgăritescu, C., Pirici, D., Simionescu, C., et al. (2012). The utility of CD44, CD117 and CD133 in identification of cancer stem cells (CSC) in oral squamous cell carcinomas (OSCC). *Romanian Journal of Morphology and Embryology, 52*, 985–993.

Méry, B., Guy, J. B., Espenel, S., et al. (2016). Targeting head and neck tumoral stem cells: From biological aspects to therapeutic perspectives. *World J Stem Cells, 8*(1), 13–21.

Misra, S., Toole, B. P., & Ghatak, S. (2006). Hyaluronan constitutively regulates activation of multiple receptor tyrosine kinases in epithelial and carcinoma cells. *The Journal of Biological Chemistry, 281*(46), 34936–34941.

Modur, V., Joshi, P., Nie, D., et al. (2016). CD24 expression may play a role as a predictive indicator and a modulator of cisplatin treatment response in head and neck squamous cellular carcinoma. *PLoS One*. https://doi.org/10.1371/journal.pone.0156651.

Noto, Z., Yoshida, T., Okabe, M., et al. (2013). CD44 and SSEA-4 positive cells in an oral cancer cell line HSC-4 possess cancer stem-like cell characteristics. *Oral Oncology, 49*(8), 787–795.

Okamoto, H., Fujishima, F., Nakamura, Y., et al. (2013). Significance of CD133 expression in esophageal squamous cell carcinoma. *World Journal of Surgical Oncology, 11*(1), 51–60.

Prince, M. E., Sivanandan, R., Kaczorowski, A., et al. (2007). Identification of a subpopulation of cells with cancer stem cell properties in head and neck squamous cell carcinoma. *Proceedings of the National Academy of Sciences of the United States of America, 104*(3), 973–978.

Qian, X., Tan, C., Wang, F., et al. (2016). Esophageal cancer stem cells and implications for future therapeutics. *OncoTargets Ther, 9*, 2247–2254.

Ren, Z. H., Zhang, C. P., & Ji, T. (2016). Expression of SOX2 in oral squamous cell carcinoma and the association with lymph node metastasis. *Oncology Letters, 11*(3), 1973–1979.

Saliba, A. E., Westermann, A. J., Gorski, S. A., et al. (2014). Single-cell RNA-seq: Advances and future challenges. *Nucleic Acids Research, 42*, 8845–8860.

Sato, N., Meijer, L., Skaltsounis, L., et al. (2004). Maintenance of pluripotency in human and mouse embryonic stem cells through activation of Wnt signaling by a pharmacological GSK-3-specific inhibitor. *Nature Medicine, 10*(1), 55–63.

Satpute, P. S., Hazarey, V., Ahmed, R., et al. (2013). Cancer stem cells in head and neck squamous cell carcinoma: A review. *Asian Pacific Journal of Cancer Prevention, 14*(10), 5579–5587.

Spiegelberg, D., Kuku, G., Selvaraju, R., et al. (2014). Characterization of CD44 variant expression in head and neck squamous cell carcinomas. *Tumor Biology, 35*(3), 2053–2062.

Takahashi-Yanaga, F., & Kahn, M. (2010). Targeting Wnt signaling: Can we safely eradicate cancer stem cells? *Clinical Cancer Research, 16*(12), 3153–3162.

Toledano, I., Graff, P., Serre, A., et al. (2012). Intensity-modulated radiotherapy in head and neck cancer: Results of the prospective study GORTEC 2004-03. *Radiotherapy and Oncology, 103*(1), 57–62.

Torre, L. A., Bray, F., Siegel, R. L., et al. (2015). Global cancer statistics, 2012. *CA: a Cancer Journal for Clinicians, 65*(2), 87–108.

Tsai, L. L., CC, Y., Chang, Y. C., et al. (2011). Markedly increased Oct4 and Nanog expression correlates with cisplatin resistance in oral squamous cell carcinoma. *Journal of Oral Pathology & Medicine, 40*(8), 621–628.

Tsai, S. T., Wang, P. J., Liou, N. J., et al. (2015). ICAM1 is a potential cancer stem cell marker of esophageal squamous cell carcinoma. *PLoS One*. https://doi.org/10.1371/journal.pone.0142834.

Valent, P., Bonnet, D., De Maria, R., et al. (2012). Cancer stem cell definitions and terminology : The devil is in the details. *Nature Review Cancer, 12*, 767–775.

Vallard, A., Espenel, S., & Guy, J. B. (2016). Targeting stem cells by radiation: From the biological angle to clinical aspects. *World Journal Stem Cells, 8*(8), 243–250.

Vasiliou, V., Vasiliou, K., & Nebert, D. W. (2009). Human ATP-binding cassette (ABC) transporter family. *Human Genomics, 3*(3), 281–290.

Vigneswaran, N., & Williams, D. M. (2014). Epidemiological trends in head and neck cancer and aids in diagnosis. *Oral and Maxillofacial Surgery Clinics of North America, 26*(2), 123–141.

Vissink, A., Mitchell, J. B., Baum, B. J., et al. (2010). Clinical management of salivary gland hypofunction and xerostomia in head and neck cancer patients: Successes and barriers. *International Journal of Radiation Oncology, Biology, Physics, 78*(4), 983–991.

Visvader, J. E., & Lindeman, G. J. (2012). Perspective cancer stem Cells : Current status and evolving complexities. *Cell Stem Cell, 10*(6), 717–728.

Vlashi, E., McBride, W., & Pajonk, F. (2009). Radiation responses of cancer stem cells. *Journal of Cellular Biochemistry, 108*(2), 339–342.

Wang, E., Casciano, C. N., Clement, R. P., et al. (2000). In vitro flow cytometry method to quantitatively assess inhibitors of P-glycoprotein. *Drug Metabolism and Disposition, 28*(5), 522–528.

Wang, J., Guo, L. P., Chen, L. Z., et al. (2007). Identification of cancer stem cell-like side population cells in human nasopharyngeal carcinoma cell line. *Cancer Research, 67*(8), 3716–3724.

Wang, S., Wong, G., & de Heer, A. (2009). CD44 variant isoforms in head and neck squamous cell carcinoma progression. *Laryngoscope, 119*(8), 1518–1530.

Wang, Y., Zhe, H., Zhang, N., et al. (2012). Cancer stem cell marker ALDH1 expression is associated with lymph node metastasis and poor survival in esophageal squamous cell carcinoma: a study from high incidence area of northern China. *Diseases of the Esophagus, 25*, 560–565.

Wicha, M., Liu, S., & Dontu, G. (2006). Cancer stem cells: An old idea – a paradigm shift. *Cancer Research, 66*(4), 1883–1890.

Yan, M., Yang, X., Wang, L., et al. (2013). Plasma membrane proteomics of tumor spheres identify CD166 as a novel marker for cancer stem-like cells in head and neck squamous cell carcinoma. *Molecular & Cellular Proteomics, 12*(11), 3271–3284.

Yang, M. H., Hsu, D. S., Wang, H. W., et al. (2010). Bmi1 is essential in Twist1-induced epithelial-mesenchymal transition. *Nature Cell Biology, 12*(10), 982–992.

Yaromina, A., Krause, M., Thames, H., et al. (2007). Pre-treatment number of clonogenic cells and their radiosensitivity are major determinants of local tumour control after fractionated irradiation. *Radiotherapy and Oncology, 83*(3), 304–310.

Yata, K., Beder, L., Tamagawa, S., et al. (2015). MicroRNA expression profiles of cancer stem cells in head and neck squamous cell carcinoma. *International Journal of Oncology, 47*(4), 1249–1256.

Yu, C., Lo, W., Chen, Y., et al. (2010). Bmi-1 regulates snail expression and promotes metastasis ability in head and neck squamous cancer-derived ALDH1 positive cells. *Journal of Oncology, 2011*, 1–16.

Zechner, D., Fuijita, Y., Hulsken, J., et al. (2003). Beta-catenin signals regulate cell growth and the balance between progenitor cell expansion and differentiation in the nervous system. *Developmental Biology, 258* (2), 406–418.

Zhang, P., Zhang, Y., Mao, L., et al. (2009). Side population in oral squamous cell carcinoma possesses tumor stem cell phenotypes. *Cancer Letters, 277*(2), 227–234.

Zhang, Q., Shi, S., Yen, Y., et al. (2010). A subpopulation of CD133+ cancer stem-like cells characterized in human oral squamous cell carcinoma confer resistance to chemotherapy. *Cancer Letters, 289*(2), 151–160.

Zhang, G., Ma, L., Xie, Y., et al. (2012a). Esophageal cancer tumorspheres involve cancer stem-like populations with elevated aldehyde dehydrogenase enzymatic activity. *Molecular Medicine Reports, 6* (3), 519–524.

Zhang, Z., Filho, M. S. A., & Nor, J. E. (2012b). The biology of head and neck cancer stem cells. *Oral Oncology, 49*(1), 1–9.

Zhu, L., Yuan, L., Wang, H., et al. (2015). A meta-analysis of concurrent chemoradiotherapy for advanced esophageal cancer. *PLoS One*. https://doi.org/10.1371/journal.pone.0128616.

Zimmerer, R., Ludwig, N., Kampmann, A., et al. (2017). CD24+ tumor-initiating cells from oral squamous cell carcinoma induce initial angiogenesis in vivo. *Microvascular Research, 112*, 101–108.

Zscheppang, K., Kurth, I., Wachtel, N., et al. (2016). Efficacy of beta1 integrin and EGFR targeting in sphere-forming human head and neck cancer cells. *Journal of Cancer, 7*(6), 736–745.

Advs Exp. Medicine, Biology - Innovations in Cancer Research and Regenerative Medicine (2018) 1: 103–124
DOI 10.1007/5584_2017_118

Published online: 15 November 2017

Contrasting Views on the Role of Mesenchymal Stromal/Stem Cells in Tumour Growth: A Systematic Review of Experimental Design

Ahmed Kolade Oloyo, Melvin Anyasi Ambele, and Michael Sean Pepper

Abstract

The effect of mesenchymal stromal/stem cells (MSCs) on tumour growth remains controversial. Experimental evidence supports both an inhibitory and a stimulatory effect. We have assessed factors responsible for the contrasting effects of MSCs on tumour growth by doing a meta-analysis of existing literature between 2000 and May 2017. We assessed 183 original research articles comprising 338 experiments. We considered (a) in vivo and in vitro experiments, (b) whether in vivo studies were syngeneic or xenogeneic, and (c) if animals were immune competent or deficient. Furthermore, the sources and types of cancer cells and MSCs were considered together with modes of cancer induction and MSC administration. 56% of

A.K. Oloyo
Institute for Cellular and Molecular Medicine, Department of Immunology, and SAMRC Extramural Unit for Stem Cell Research and Therapy, Faculty of Health Sciences, University of Pretoria, Pretoria, South Africa

Department of Physiology, Faculty of Basic Medical Science, College of Medicine, University of Lagos, Lagos, Nigeria
e-mail: akoloyo@cmul.edu.ng

M.A. Ambele
Institute for Cellular and Molecular Medicine, Department of Immunology, and SAMRC Extramural Unit for Stem Cell Research and Therapy, Faculty of Health Sciences, University of Pretoria, Pretoria, South Africa

Department of Oral Pathology and Oral Biology, School of Dentistry, Faculty of Health Sciences, University of Pretoria, Pretoria, South Africa
e-mail: ambelema@gmail.com

M.S. Pepper (✉)
Institute for Cellular and Molecular Medicine, Department of Immunology, and SAMRC Extramural Unit for Stem Cell Research and Therapy, Faculty of Health Sciences, University of Pretoria, Pretoria, South Africa
e-mail: michael.pepper@up.ac.za

all 338 experiments reported that MSCs promote tumour growth. 78% and 79% of all experiments sourced human MSCs and cancer cells, respectively. MSCs were used in their naïve and engineered form in 86% and 14% of experiments, respectively, the latter to produce factors that could alter either their activity or that of the tumour. 53% of all experiments were conducted in vitro with 60% exposing cancer cells to MSCs via coculture. Of all in vivo experiments, 79% were xenogeneic and 63% were conducted in immune-competent animals. Tumour growth was inhibited in 80% of experiments that used umbilical cord-derived MSCs, whereas tumour growth was promoted in 64% and 57% of experiments that used bone marrow- and adipose tissue-derived MSCs, respectively. This contrasting effect of MSCs on tumour growth observed under different experimental conditions may reflect differences in experimental design. This analysis calls for careful consideration of experimental design given the large number of MSC clinical trials currently underway.

Keywords
Cancer • Mesenchymal stem cell • Syngeneic • Tumour • Xenogeneic

Abbreviations

AD	Adipose tissue
BM	Bone marrow
CC	Coculture
CCR2	C-C motif chemokine receptor 2
c-Kit	Tyrosine-protein kinase Kit also known as mast/stem cell growth factor receptor (SCFR)
CM	Conditioned medium
c-Met	Tyrosine-protein kinase Met or hepatocyte growth factor receptor
CXCR4	C-X-C motif chemokine receptor 4
EGF	Epithelial growth factor
HGF	Hepatocyte growth factor
HNSCC	Head and neck squamous cell carcinoma
IL-1β	Interleukin 1-beta
IL-8	Interleukin 8
IP	Intraperitoneal
IV	Intravenous
MCP-1	Monocyte chemotactic protein 1
MSC	Mesenchymal stromal/stem cell
PDGF	Platelet-derived growth factor
SC	Subcutaneous
SCF	Stem cell factor
SCID	Severe combined immunodeficiency
SDF-1	Stromal cell-derived factor 1
TGF-ß	Transforming growth factor-beta
TNF-α	Tumour necrosis factor alpha
UC	Umbilical cord
VEGF	Vascular endothelial growth factor
VEGFR	Vascular endothelial growth factor receptor

1 Introduction

Interest in the effect of mesenchymal stromal/stem cells (MSCs) on tumour growth stems from two areas. The first relates to the fact that MSCs are being assessed in a growing number of clinical trials for a wide variety of diseases (Hong et al. 2014; Squillaro et al. 2016). The fear is that systemically administered MSCs have the potential to activate dormant tumours through the production of paracrine growth stimulatory molecules (Lazennec and Lam 2016). The second relates to the fact that in some experimental settings, MSCs have been shown to inhibit tumour growth, and this has sparked interest in the possible use of MSCs in the treatment of cancer.

Globally, cancer remains a leading cause of death. Cancer incidence and cancer-related mortality increased by approximately 11% and 17%, respectively, between 2008 and 2012. This trend is projected to increase by about 70% in the next two decades, with cancer incidence increasing from 14.1 million in 2012 to 22 million in 2030, while mortality will increase from 8.2 to 13 million (Ferlay et al. 2010). In 2008, about 169.3 million years of healthy life were lost due to cancer (Soerjomataram et al. 2012). While primary prevention of cancer includes raising public awareness and avoiding modifiable risk factors, there is a need for effective treatment for those already afflicted.

Several therapeutic measures exist for cancer, including chemotherapy, radiotherapy and immunotherapy. These therapies have their own side effects and limitations. Recently, the concept of cellular therapy for cancer was introduced, even though the effect of stem cell treatment on cancer is highly controversial (Hong et al. 2014). Mesenchymal stromal/stem cells (MSCs) contain cells with stem cell-like properties that are multipotent in nature and are able to self-renew (Bianco et al. 2013). It has also been reported that they have the ability to home to sites of injury and inflammation and to tumours (Hong et al. 2014). The therapeutic potential of MSCs may lie in cellular rejuvenation or as a transport vehicle for other therapies (Serakinci et al. 2014). Hong, Lee and Kang provide a detailed explanation of the different interactions between MSCs and tumours (Hong et al. 2014). Here we have assessed whether there is a relationship between experimental design and observed results.

MSCs on their own are believed not to be tumorigenic, but several studies have reported both tumour-promoting (Albarenque et al. 2011; De Boeck et al. 2013; Ljujic et al. 2013; Zhang et al. 2013) and inhibitory (Chao et al. 2012; Ganta et al. 2009; Maurya et al. 2010) effects. Experimental design is highly variable. In vivo experiments may be xenogeneic, syngeneic or isogeneic. The immune status of the animal may be immune competent, compromised or deficient. Outcomes of in vitro experiments could be influenced by whether MSCs and cancer cells were cocultured or cancer cells were exposed to conditioned media from MSCs. The sources and types of cancer cells and MSCs may influence the outcome of the experiments. MSCs can be sourced from different animals including rabbits, mice and humans and can be found in various tissues including bone marrow, umbilical cord blood, peripheral blood, placenta and adipose tissue. Experimental design may therefore have an important influence on the outcomes of experiments that assess the tumorigenic action of MSCs.

Understanding how MSCs interact with cancer cells and the experimental factors that influence the results may direct future research and the ultimate use of MSCs to treat cancer. Likewise, the incidental tumour-promoting effects of MSCs on latent/dormant tumours in patients being treated for other conditions need to be avoided. This is because tumour microenvironment continuously produces and releases various cytokines and mediators that establish a state of inflammation which has the capacity to attract MSCs. This tumour-directed migratory potential of MSCs has been observed in almost all cancer types tested so far which includes breast (Patel et al. 2010), lung (Loebinger et al. 2009), ovarian (Kidd et al. 2009), pancreatic (Zischek et al. 2009), colon (Menon et al. 2007), skin (Studeny et al. 2002) and brain cancer (Sasportas et al. 2009), even though the underlying mechanism of this MSC tropism remains unknown. Stem cell factor (SCF)/c-Kit, SDF-1/CXCR4, VEGF/VEGFR, HGF/c-Met and MCP-1/CCR2 are some of the chemokine/receptor pairs reported to be associated with homing of MSCs to disease sites. In addition, TGF-ß, IL-8, EGF, neurotrophin-3, TNF-α, PDGF and IL-1ß are other growth factors, angiogenic factors and inflammatory cytokines known to stimulate MSC migration. Most of these chemokines and cytokines are produced and released by tumours (Motaln et al. 2010; Nakamizo et al. 2005), which may serve as chemoattractants (ligands) for receptors on MSCs. This chemokine/receptor axis between tumours and MSCs may lead MSCs that are administered to patients for the treatment of other diseases migrating and homing to sites of latent/dormant tumours, thereby stimulating their growth.

Here we have reviewed available published literature over the last 16 years which has assessed the effects of MSCs on tumour growth. We (a) looked at which experimental factors were associated with specific outcomes and (b) how these factors might have influenced experimental outcomes.

2 Methods

We conducted a systematic review and a meta-analysis of the available literature from January 2000 to May 2017. We used the search terms MSC, cancer and tumour growth on Google Scholar and PubMed search engines. A total of 1586 articles were generated from which we selected 183 after applying our exclusion criteria. These 183 articles comprised 338 experiments that assessed the effects of MSCs on tumorigenesis.

2.1 Inclusion Criteria

We included original research articles published in or with an expanded abstract in English between January 2000 and May 2017. The earliest article testing the effect of MSCs on tumour progression was published in 2003 (Djouad et al. 2003). All included articles have a definite end point regarding the effect of MSCs on tumour growth or metastasis.

2.2 Exclusion Criteria

Duplicate and non-original research articles, such as review articles, were excluded. Articles that studied the effect of MSCs on pathologies other than cancer/tumours were excluded. Articles that studied the effect of other substances besides MSCs on cancer were excluded. We excluded studies where no definite effects of MSCs on tumour progression were reported. Studies where MSCs were derived from tumours or other pathological tissues were also excluded (Fig. 1).

Search terms

("mesenchymal stromal cells"[MeSH Terms] OR ("mesenchymal"[All fields] AND "Stromal"[All fields] AND "cells"[All fields]) OR "mesenchymal stromal cells"[All fields) AND ("mesenchymal stromal cells"[MeSHTerms] OR ("mesenchymal "[ALL fields] AND "stromal"[Al fields] AND "cells"[All fields]) OR "mesenchymal stromal cells"[All fields] OR ("mesenchymal"[All fields] AND "stem cells"[All fields]) OR "mesenchymal stem cells"[All fields]) AND ("neoplasms"[MeSH Terms] OR "neoplasms"[All fields] OR"cancer"[All field])

↓

Articles generated

n=1586

↓

Exclusion criteria

Studies with no definite effect of MSCs on tumour

Articles where the effect of MSCs was not studied on cancer

Studies on effect of other substances besides MSCs on cancer

Studies using tumour-derived MSCs

Review articles

↓

Articles selected for study

n=183

Fig. 1 Method of searching the literature for the effect of MSCs on tumour growth

3 Results and Discussion

3.1 Effects of MSCs on Tumour Growth (Inhibition Versus Stimulation)

Our review revealed that MSCs had a stimulatory effect on tumour growth in 56% (90 in vivo and 100 in vitro experiments) and an inhibitory effect in 44% (69 in vivo and 79 in vitro experiments) of all studies assessed (Fig. 2). The response of tumours to MSCs was not evenly distributed per experimental type, exposure type, experimental animals used, MSCs or cancer cell types.

The effects (stimulatory or inhibitory) of different MSC factors/parameters considered in this review on tumour growth in vivo or in vitro are summarized in Table 1.

3.2 Types of Experiment (In Vivo Versus In Vitro)

One hundred seventy-nine (53%) of the 338 experiments reviewed were conducted in vitro, of which 100 (56%) reported a stimulatory effect on tumour growth (Fig. 3). Forty-seven per cent (159) of experiments were conducted in vivo (Fig. 3), of which 90 (57%) revealed that MSCs promote tumour growth (Fig. 3). The secretome of transplanted MSCs is known to be largely determined by their microenvironment, and the same MSCs will have a different profile in vitro to that in in vivo when they are transplanted (Dittmer and Leyh 2014). The lack of differentiation between tumour response and experimental type indicates a need to conduct simultaneous in vivo and in vitro experiments and to interpret the latter with particular caution.

3.3 Effect of MSCs on Tumour Growth: The Role of In Vivo Specific Factors

The effect of the immune status of the animal and the nature of the animal model and experimental design (syngeneic or xenogeneic) are some of the in vivo parameters/factors which are likely to affect the outcome of studies on the effect of MSCs on tumour growth.

3.3.1 Immune Status of Experimental Animals

One hundred one (64%) of the 159 in vivo experiments used immune-competent animals, while 58 (36%) used immune-deficient or immune-compromised animals. Of the 159 in vivo experiments reviewed, 37 used severe combined immunodeficiency (SCID) or athymic mice in a xenogeneic experimental design. Quante et al. (2011) is the only syngeneic experimental study in SCID mice that assessed the effect of murine BM-MSCs on mouse lung cancer, and this revealed a stimulatory effect

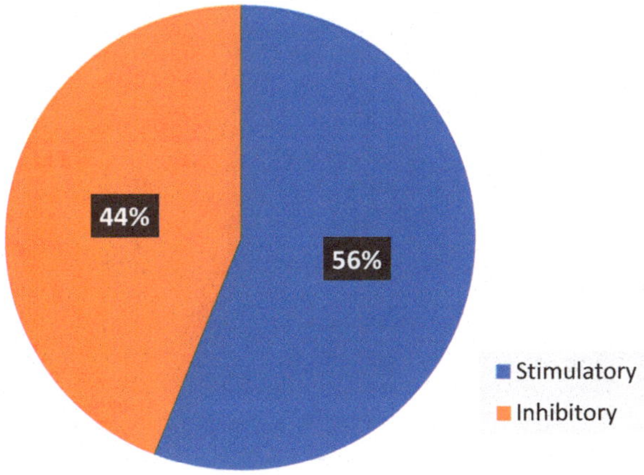

Fig. 2 The effect of MSCs on tumour growth

Table 1 The effect of MSCs on tumour growth

Experimental type ($n = 338$)	In vivo ($n = 159$; 47%)		In vitro ($n = 179$; 53%)	
Effect on tumour growth	*Stimulatory*	*Inhibitory*	*Stimulatory*	*Inhibitory*
	$n = 90$ (57%)	$n = 69$ (43%)	$n = 100$ (56%)	$n = 79$ (44%)
Experimental model/ design	Syngeneic ($n = 22$)	Syngeneic ($n = 15$)	n/a	n/a
	Xenogeneic ($n = 68$)	Xenogeneic ($n = 54$)		
Animal model	Mouse ($n = 87$)	Mouse ($n = 61$)	n/a	n/a
	Rat ($n = 2$)	Rat ($n = 7$)		
	Other ($n = 1$)	Other ($n = 1$)		
Animal immune status	Competent ($n = 62$)	Competent ($n = 39$) Deficient and compromised ($n = 30$)	n/a	n/a
	Deficient and compromised ($n = 28$)			
Species from which MSCs were derived	Human ($n = 65$)	Human ($n = 49$)	Human ($n = 80$)	Human ($n = 68$)
	Mouse ($n = 22$)	Mouse ($n = 11$)		
	Rat ($n = 3$)	Rat ($n = 8$)	Mouse ($n = 15$)	Mouse ($n = 5$)
		Hamster ($n = 1$)	Rat ($n = 5$)	Rat ($n = 6$)
Source of MSCs	BM ($n = 67$)	BM ($n = 40$)	BM ($n = 65$)	BM ($n = 34$)
	AD ($n = 10$)	AD ($n = 9$)	AD ($n = 23$)	AD ($n = 16$)
	UC ($n = 6$)	UC ($n = 15$)	UC ($n = 4$)	UC ($n = 25$)
	Others ($n = 7$)	Others ($n = 4$)	Others ($n = 8$)	Others ($n = 4$)
Sources of cancer cells	Human ($n = 65$)	Human ($n = 47$)	Human ($n = 88$)	Human ($n = 66$)
	Mouse ($n = 22$)	Mouse ($n = 13$)		
	Rat ($n = 2$)	Rat ($n = 6$)	Mouse ($n = 11$)	Mouse ($n = 8$)
	Chemical ($n = 1$)	Chemical ($n = 3$)	Rat ($n = 1$)	Rat ($n = 4$)
				Chemical ($n = 1$)
Types of cancer	Breast ($n = 22$)	Breast ($n = 14$)	Breast ($n = 36$)	Breast ($n = 24$)
	Lung ($n = 7$)	Lung ($n = 8$)	Lung ($n = 5$)	Lung ($n = 8$)
	Colorectal ($n = 14$)	Colorectal ($n = 2$)	Colorectal ($n = 5$)	Colorectal ($n = 4$)
		Prostate ($n = 9$)		
	Prostate ($n = 7$)	Glioma ($n = 10$)	Prostate ($n = 11$)	Prostate ($n = 4$)
	Glioma ($n = 3$)	HNSCC ($n = 1$)		Glioma ($n = 9$)
	HNSCC ($n = 2$)	Hepatic ($n = 7$)	Glioma ($n = 3$)	HNSCC ($n = 3$)
	Hepatic ($n = 1$)	Gastric ($n = 1$)	HNSCC ($n = 6$)	
	Gastric ($n = 9$)	Sarcoma ($n = 4$)	Hepatic ($n = 5$)	Hepatic ($n = 7$)
	Sarcoma ($n = 9$)	Others ($n = 13$)	Gastric ($n = 7$)	Gastric ($n = 2$)
	Others ($n = 16$)		Sarcoma ($n = 7$)	Sarcoma ($n = 4$)
			Others ($n = 15$)	Others ($n = 14$)
Methods of cancer induction	SC ($n = 58$)	SC ($n = 27$)		
	IV ($n = 4$)	IV ($n = 11$)	Coculture (n = 59)	Coculture (n = 46)
	IP ($n = 4$)	IP ($n = 6$)		
	Ortho ($n = 19$)	Ortho ($n = 17$)	Conditioned medium (n = 41)	Conditioned medium (n = 33)
	Others ($n = 5$)	Others ($n = 8$)		
Mode of administration of MSCs	SC ($n = 54$)	SC ($n = 17$)		
	IV ($n = 14$)	IV ($n = 25$)		
	IP ($n = 4$)	IP ($n = 9$)		
	Intra-tumoural ($n = 13$)	Intra-tumoural ($n = 11$)		
	Others ($n = 5$)	Others ($n = 7$)		

(continued)

Table 1 (continued)

Experimental type ($n = 338$)	In vivo ($n = 159$; 47%)		In vitro ($n = 179$; 53%)	
MSCs status	Naïve ($n = 84$)	Naïve ($n = 44$)	Naïve ($n = 95$)	Naïve ($n = 66$)
	Engineered ($n = 6$)	Engineered ($n = 25$)	Engineered ($n = 5$)	Engineered ($n = 13$)

n number of studies, *SC* subcutaneous, *IV* intravenous, *IP* intraperitoneal, *Ortho* orthotopically, *HNSCC* head and neck squamous cell carcinoma, *n/a* not applicable

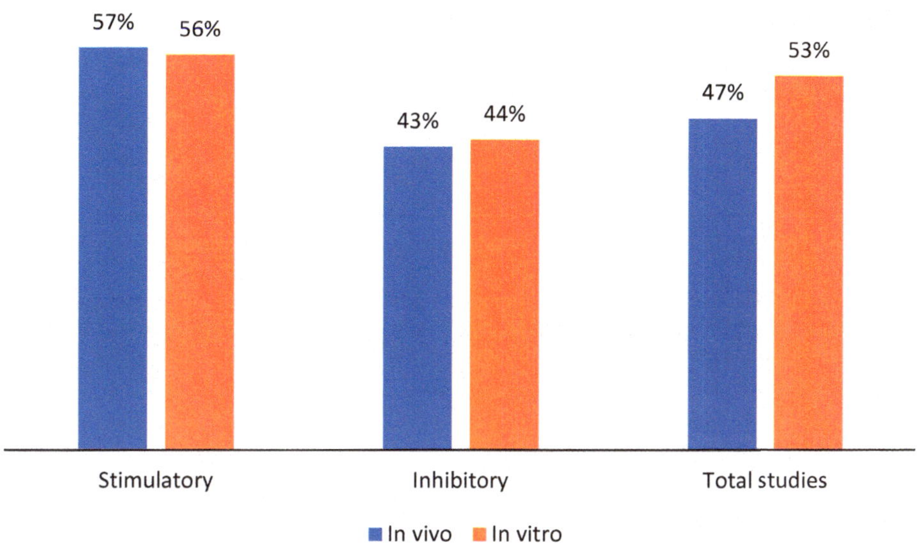

Fig. 3 Experimental type (in vivo and in vitro) used to assess the effect of MSCs on tumour growth. Virtually equal numbers of studies showed stimulatory or inhibitory effects although the number of studies conducted in vitro was slightly higher

(Quante et al. 2011). Conducting xenogeneic experiments using immune-deficient animals may reduce the immune response of the host to both the cancer and MSCs from other species. Immune-competent animals with intact immunosurveillance systems should have a natural resistance to and reject either or both cancer cells and MSCs from another species.

MSCs stimulated tumour growth in 61% ($n = 62$) of experiments that used immune-competent animals, suggesting an interaction with the host immune system. MSCs inhibited tumour growth in 52% ($n = 30$) of experiments that used immune-deficient or immune-compromised animals (Fig. 4). Immune-deficient animals such as athymic mice have been used to validate human MSCs prior to phase II clinical trials. Even though human cells are successfully transplanted into these animals and subsequently survive and thrive in them, the lack of a competent immune system can mask natural responses to MSCs (Tholpady et al. 2003) and tumour cells. Athymic animals are also prone to developing subclinical infections and systemic illness (Lopez and Spencer 2011), which may mask the effect of MSCs. The immune status of animals used for in vivo experiments is therefore likely to play an important role in determining the effect of MSCs on tumour growth.

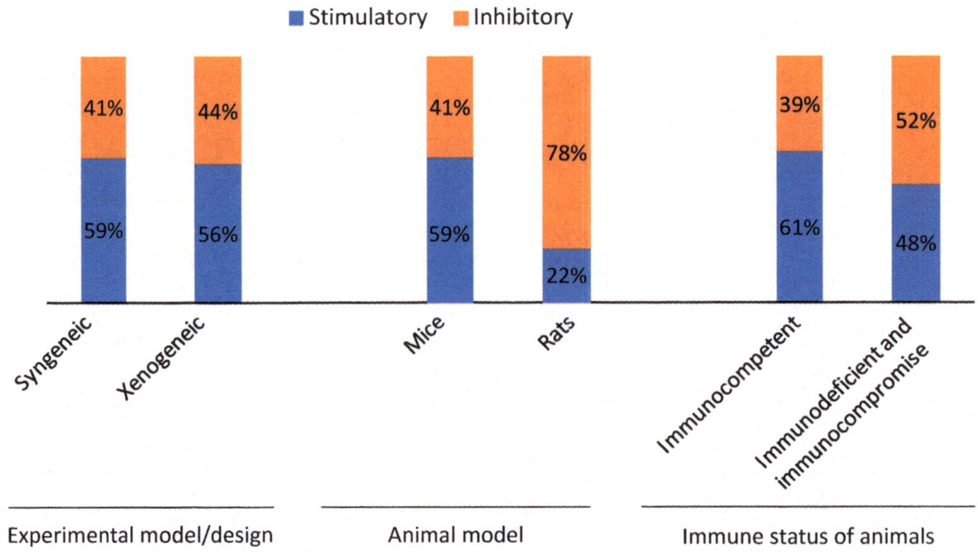

Fig. 4 Effect of in vivo specific factors on tumour growth in response to administered MSCs. A greater percentage of studies showed that MSCs promote tumour growth in vivo in mice and immune-competent animals, whereas they inhibit tumour growth in rats and immune-deficient animals

3.3.2 Species in Experimental Animal Models

One hundred forty-eight (93%) of the in vivo experiments used mice, while nine used rats (6%) and other models including hamster and rabbit (1%; $n = 2$). MSCs stimulated tumour growth in 59% ($n = 87$) of in vivo experiments using mice, whereas tumour growth was inhibited in 78% ($n = 7$) of studies using rats (Fig. 4), although the number of experiments using rats ($n = 9$) was very small compared to mice ($n = 148$).

3.3.3 Experimental Model/Design

The source of MSCs and cancer cell lines used for in vivo studies was mainly human. Most of the in vivo experiments – 122 (77%) – were xenogeneic, while the remaining 37 (23%) were syngeneic. MSCs promoted tumour growth in 56% ($n = 68$) and 59% ($n = 22$) of xenogeneic and syngeneic studies, respectively (Fig. 4). The origin of MSCs and cancer cells may affect the immune response in the experimental animals employed, and differences have been reported between allogenic and xenogeneic experiments in several animal models (Revell and Athanasiou 2009; Sigrist et al. 2005).

3.4 MSC Sources and Types Used in In Vivo and In Vitro Experiments

MSCs used were from humans (78%; $n = 262$), mice (16%; $n = 53$), rats (6%; $n = 22$) and hamsters ($n = 1$). Tumour growth was stimulated in 55% ($n = 145$) and 70% ($n = 37$) of experiments that used MSCs from humans and mice, respectively, whereas growth was inhibited in 64% ($n = 14$) of experiments that used rat MSCs (Fig. 5). Sources of MSCs may influence the immune response of the animals used in in vivo experiments. Using xenogeneic or syngeneic cells in in vivo experiments may affect the immune system (Fig. 4) and thus influence the effect of MSCs on tumour growth.

MSCs were derived from BM, umbilical cord (UC) and adipose tissue (AD), and a few studies used MSCs derived from peripheral blood

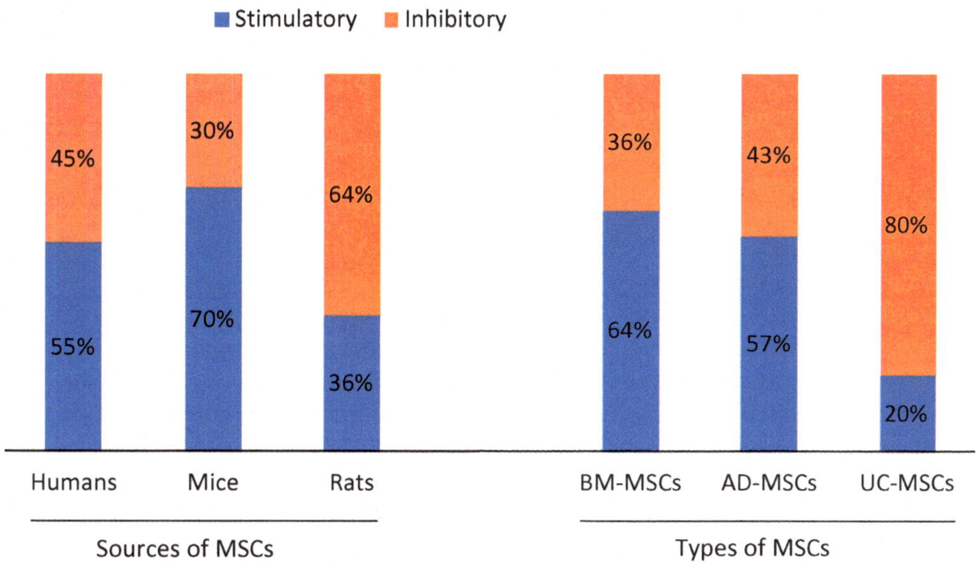

Fig. 5 Effect of sources and types of MSCs on tumour growth. A greater proportion of studies analysed showed that MSCs from humans and mice promote tumour growth, while rat MSCs showed an inhibitory effect regardless of experimental type. BM- and AD-MSCs promote tumour growth in most of the studies where they were used unlike UC-MSCs which inhibited tumour growth irrespective of the experimental type

mononuclear cells, foetal dermis, liver and umbilical cord blood, amongst others. Of experiments 61% ($n = 206$) used BM-MSCs, 15% ($n = 50$) used UC-MSCs, 17% ($n = 58$) used AD-MSCs, while the remaining 7% ($n = 23$) were sourced from other tissues like dermis, decidua, liver, umbilical cord blood and peripheral blood. BM-MSCs stimulated tumour growth in 64% ($n = 132$) of experiments (Fig. 5), regardless of whether the experiment was conducted in vivo or in vitro. BM-MSCs stimulated tumour growth in 66% ($n = 65$) of in vitro experiments and 63% ($n = 67$) of in vivo experiments. The stimulatory effect of BM-MSCs was primarily associated with breast cancer cells (Supplementary Tables S1a, S1b and S2a, S2b). Studies assessing the action of BM-MSCs on breast cancer cells were highly prevalent amongst those reviewed. UC-MSCs inhibited tumour growth in 80% ($n = 40$) of experiments where they were used (Fig. 5) regardless of the experimental type. Experiments were conducted both in vitro ($n = 29$) and in vivo ($n = 21$). UC-MSCs inhibited tumour growth in 86% ($n = 25$) of in vitro experiments and in 71% ($n = 15$) of in vivo experiments (Supplementary Tables S3a, S3b and S4a, S4b). Tumour growth was promoted in 57% ($n = 33$) of studies where AD-MSCs were used irrespective of the experimental type (Fig. 5). Experiments were conducted both in vitro ($n = 39$) and in vivo ($n = 19$). AD-MSCs promoted tumour growth in 53% ($n = 10$) of in vivo and in 59% ($n = 23$) of in vitro experiments (Supplementary Tables S5a, S5b and S6a, S6b). MSCs derived from other tissue sources such as dermis, peripheral blood and umbilical cord blood had a stimulatory (65%; $n = 15$) or inhibitory (35%; $n = 8$) effect on tumour growth.

Even though MSCs isolated from distinct tissue sources display some characteristics that are similar, certain inherent genetic or cellular variations exist between tissues (Wagner et al. 2005; Zhou et al. 2013). For example, breast cancer may be stimulated by BM-MSCs and inhibited by UC-MSCs, or AD-MSCs may inhibit prostate cancer but promote melanomas (Supplementary Tables S5a, S5b and S6a, S6b). It thus appears that the type of MSCs used is an important factor that influences tumour growth in vivo and in vitro.

3.5 Status of MSCs Used in Experimental Studies

MSCs were used either in their natural form after expansion, or they were modified or genetically altered to produce a particular cytokine or chemokine. MSCs used in their native form after expansion are referred to as naïve MSCs, and modified/altered MSCs that produce tailor-made effects are referred to as engineered MSCs. In this review, 289 (85%) of studies used naïve MSCs (Table 1). 179 (62%) studies reported a stimulatory effect on tumour growth by naïve MSCs. Tumour growth was inhibited in 38 (78%) experiments where engineered MSCs were used (Table 1). The inhibitory effect of engineered MSCs on tumour growth is not surprising, given that these MSCs were engineered to produce substances that are known to possess tumouricidal or tumour growth inhibitory properties (Li et al. 2014; Nakamura et al. 2004).

3.6 Cancer Sources and Types Used to Evaluate the Effect of MSCs on Tumour Growth

Two hundred sixty-six (79%) of the 338 experiments analysed used human cancer cells, 16% ($n = 54$) used murine cancer cells, 4% ($n = 13$) used rat cancer cells and 1% ($n = 5$) of the experiments induced cancer using chemical methods. MSCs promoted growth of human and mouse cancer cells in 153 (57%) and 33 (61%) of studies, respectively, whereas MSCs inhibited growth of rat cancer cells in 10 (77%) studies (Fig. 6). Tumour growth was inhibited in both experiments in which cancer was induced by chemical means (Chen et al. 2014b; Paris et al. 2016).

The effects of MSCs on breast cancer were studied in 96 (29%) of the experiments included in this review. The effects of MSCs on lung cancer (8%; $n = 28$), prostate cancer (9%; $n = 31$), glioma (7%; $n = 25$), colorectal

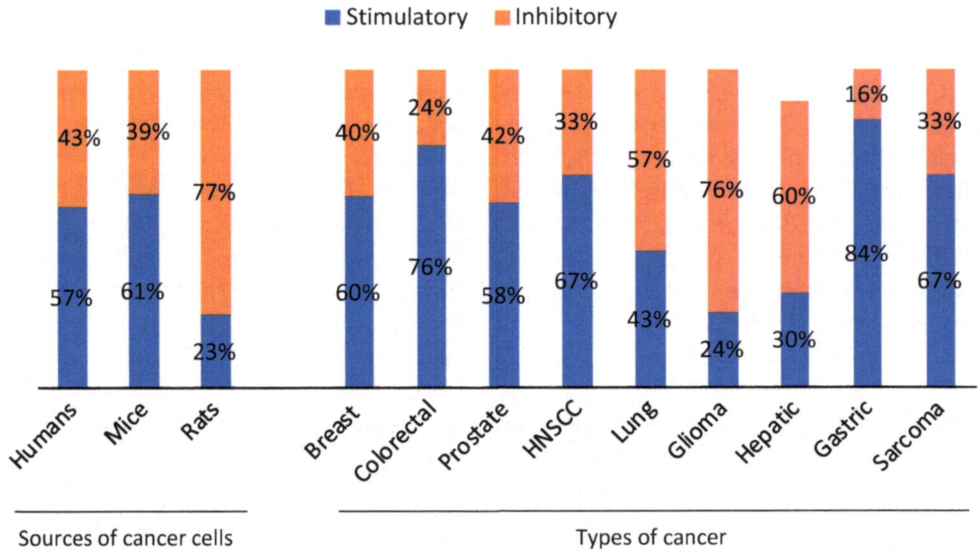

Fig. 6 Effect of MSCs on the sources and types of cancer cells studied in vivo and in vitro. The majority of the studies using cancer cells from humans and mice revealed that growth was promoted by MSCs, while growth of cancer cells from rats was inhibited by MSCs in the majority of studies. MSCs promoted growth of breast, colorectal, prostate and gastric cancers, HNSCC and sarcoma in the majority of the studies in which they were used, whereas an inhibitory effect of MSCs on lung, hepatic and glioma tumour growth was observed in the majority of the studies in which they were used

carcinoma (7%; $n = 25$), HNSCC (4%; $n = 12$), hepatic cancer (6%; $n = 20$), gastric cancer (6%; $n = 19$), sarcoma (7%; $n = 24$) and others (17%; $n = 58$) were studied in experiments included in this review. Cancer types classified as others include melanoma, myeloma, pancreatic cancer, cancer of the bladder, lymphoma, and ovarian cancer, amongst others. Different types of cancer displayed different susceptibility to MSCs in vivo and in vitro. For instance, MSCs stimulated the growth of breast cancer in 60% ($n = 58$), colorectal cancer in 76% ($n = 19$), prostate cancer in 58% ($n = 18$), gastric cancer in 84% ($n = 16$), sarcoma in 67% ($n = 16$) and HNSCC in 67% ($n = 8$) of experiments in which they were used. Conversely, MSCs inhibited lung cancer in 57% ($n = 16$), hepatic cancer in 60% ($n = 14$) and glioma in 76% ($n = 19$) of experiments in which they were studied (Fig. 6). Studies carried out on breast cancer used BM-MSCs (47%; $n = 45$), UC-MSCs (22%; $n = 21$) and AT-MSCs (19%; $n = 18$).

3.7 Methods of Inducing Cancer and Administering MSCs In Vivo

Most of the in vivo experiments included in this review used a first-generation mouse model for human cancer involving xenogeneic or syngeneic transplants (Bock et al. 2014). Tumour cells were implanted subcutaneously or orthotopically into the experimental animal. In 85 (53%) of the in vivo experiments, cancer cells were injected subcutaneously. Cancer cells were injected orthotopically (23%; $n = 36$), intravenously (9%; $n = 15$), intraperitoneally (6%; $n = 10$) or via other routes (8%; $n = 13$) in the remaining in vivo experiments.

MSCs exhibited a stimulatory effect on tumour growth in 68% ($n = 58$) and 53% ($n = 19$) of in vivo experiments where cancer cells were transplanted subcutaneously or orthotopically, respectively. Conversely, tumour growth was inhibited by MSCs in experiments where cancer cells were transplanted intravenously (73%; $n = 11$), intraperitoneally (60%; $n = 6$) or via other routes (61%; $n = 8$) (Fig. 7).

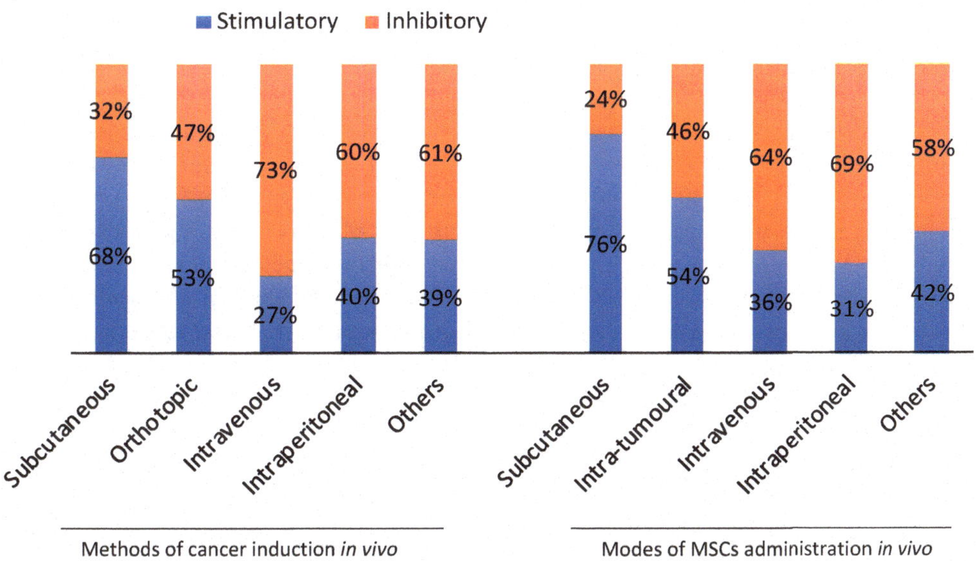

Fig. 7 Effect of the methods of cancer induction and MSC administration in vivo on tumour growth. The majority of experiments showed a stimulatory effect on tumour growth by MSCs either when the tumour was induced or the MSCs were administered subcutaneously or orthotopically, whereas tumour growth was inhibited in the majority of studies in which the cancer was induced or MSCs were administered intravenously, intraperitoneally or via other methods

The ability of MSCs to migrate to tumour sites (tumour tropism) is one of the features purportedly associated with MSCs therapy. MSCs are known to reach tumours via the vascular system. In in vivo experiments, MSCs were administered subcutaneously (45%; $n = 71$), intravenously (24%; $n = 39$), intraperitoneally (8%; $n = 13$) and via intra-tumoural injection (15%; $n = 24$). Other studies (8%; $n = 12$) administered MSCs via intramuscular and intraarterial routes. Tumour growth was promoted in 54 (76%) and in 13 (54%) experiments where MSCs were administered subcutaneously and intra-tumourally, respectively. Conversely, tumour growth was inhibited in 25 (64%), 13 (69%) and 7 (58%) of experiments that administered MSCs intravenously, intraperitoneally or via other routes, respectively (Fig. 7). The route of MSC administration appears to determine access to the tumour which is likely to determine if MSCs will be able to interact directly with the tumour.

3.8 Methods of Exposure of Cancer to MSCs In Vitro

To assess the effect of MSCs on tumour growth in in vitro experiments, cancer cells were either cocultured with MSCs or they were exposed to MSC conditioned medium. Cancer cells and MSCs were cocultured in 105 (59%) of the in vitro experiments, while cancer cells were exposed to MSC conditioned media in 74 (41%) of the in vitro experiments. Cancer growth was stimulated by MSCs in in vitro experiments either when they were cocultured with MSCs (56%; $n = 59$) or when the cancer cells were exposed to MSC conditioned medium (55%; $n = 41$). Exposure of MSCs to cancer cells via coculture experiments or conditioned medium may affect the growth of tumour cells differently in in vitro experiments. In coculture experiments, cytokines and/or chemokines from MSCs diffuse towards and influence the activities of cancer cells, while secretions from cancer cells also diffuse towards and influence the activity of MSCs. Conversely, in experiments where cancer cells are exposed to MSC conditioned media, only secretions (cytokines and/or chemokines) from MSCs in the conditioned media will influence the activity of cancer cells and not vice versa.

4 Conclusions

Our review of original research articles assessing the effect of MSCs on tumour growth has revealed the existence of varied responses to MSCs, which may be due to several experimental factors such as the origin of the MSCs and cancer cells, the route of administration of MSCs, methods of inducing cancer and the immune status of the experimental animals as well as the experimental animal model used. The diversity of experimental factors greatly limits the interpretation and comparison of different studies even when performed under similar conditions. However, we have attempted to summarize our assessment of the above factors in the 338 experimental studies reviewed and have only considered those experimental factors for which the number of in vivo and in vitro experiments is ≥ 10 and the difference in the effect on tumour growth by MSCs is $\geq 10\%$. This analysis is shown in Table 2.

In summary, the administration of MSCs or induction of cancer in in vivo experiments via subcutaneous injection stimulated tumour growth, whereas tumour growth was inhibited when these procedures were done intraperitoneally or intravenously. Both coculture and exposure of tumour cells to MSC condition medium in vitro stimulated tumour growth.

When MSCs or cancer cells from mouse were used, this resulted in an overall stimulatory effect on mouse and human tumour cell growth in both in vivo and in vitro experiments. MSCs or cancer cells from human showed an overall stimulatory effect on tumour growth in vivo, whereas in in vitro experiments a stimulatory effect was observed only when cancer cells from human were used. MSCs from rat showed an overall inhibitory effect on tumour growth in both in vivo and in vitro experiments.

Table 2 Summary of some of the experimental factors which are likely to have affected the outcome of the studies assessed. Only factors with ≥10 experimental studies in both in vivo and in vitro settings and for which the difference in experimental outcome was ≥10% were selected

Effect on tumour growth	Stimulatory		Inhibitory	
Experimental condition	In vivo	In vitro	In vivo	In vitro
Mode of MSC administration	SC ($n = 71$; 76%)	CC ($n = 105$; 56%) and CM ($n = 74$; 55%)	IP ($n = 13$; 69%) and IV ($n = 39$; 64%)	n/a
Method of cancer induction	SC ($n = 85$; 68%)		IP ($n = 10$; 60%) and IV ($n = 15$; 73%)	
Source of MSCs	Mouse ($n = 33$; 67%) and human ($n = 114$; 57%)	Mouse ($n = 20$; 75%)	Rat ($n = 11$; 73%)	Rat ($n = 11$; 55%)
Source of cancer cells	Human ($n = 112$; 58%) and mouse ($n = 35$; 63%)	Human ($n = 154$; 57%) and mouse ($n = 19$; 58%)		
Origin of MSCs	BM ($n = 107$; 63%)	BM ($n = 99$; 66%) and AD ($n = 39$; 59%)	UC ($n = 21$; 71%)	UC ($n = 29$; 86%)
Immune status of animal	Immune competent ($n = 101$; 61%)	n/a		n/a
Animal model	Mouse ($n = 148$; 59%)			
Experimental design	Syngeneic ($n = 37$; 59%) and xenogeneic ($n = 112$; 56%)			
Type of cancer	Breast ($n = 36$; 61%), sarcoma ($n = 13$; 69%), colorectal ($n = 16$; 87%) and gastric ($n = 10$; 90%)	Breast ($n = 60$; 60%), sarcoma ($n = 11$; 64%) and prostate ($n = 15$; 73%)	Prostate ($n = 16$; 56%) and glioma ($n = 13$; 77%)	Glioma ($n = 12$; 75%), lung ($n = 13$; 62%) and hepatic ($n = 12$; 58%)

SC subcutaneous, *IV* intravenous, *IP* intraperitoneal, *CC* coculture, *CM* conditioned medium, *BM* bone marrow, *AD* adipose derived, *UC* umbilical cord, *NA* not applicable, *n* number of experiments, *n/a* not applicable

In both in vivo and in vitro experiments, BM-MSCs showed a stimulatory effect on tumour growth, while an inhibitory effect was seen in response to UC-MSCs. AD-MSCs showed a stimulatory effect on tumour growth only in in vitro experiments. In cases where immune-competent animals were used and when the experimental animal was mouse, irrespective of whether the model was syngeneic or xenogeneic in design, there was an overall stimulatory effect of MSCs on tumour growth in vivo.

MSCs stimulated tumour growth in both in vivo and in vitro experiments in which breast cancer and sarcoma were used, whereas a stimulatory effect of MSCs on colorectal and gastric cancer was only observed in in vivo experiments. An overall inhibitory effect on tumour growth by MSCs was observed in glioma, whereas growth of lung and hepatic cancer was inhibited by MSCs in in vitro experiments only. Experiments on prostate cancer showed the opposite effect in vivo and in vitro as an overall stimulatory and inhibitory effect was observed in the former and the latter, respectively.

It is believed that MSCs have the ability to migrate and engraft at tumour sites where they either exert a stimulatory or inhibitory effect on tumour growth (Hong et al. 2014; Kidd et al. 2009; Lazennec and Lam 2016; Ridge et al. 2017). How the tumour cells and MSCs interact or crosstalk with each other (directly or indirectly) will determine if MSCs will either stimulate or inhibit tumour growth. MSCs are known to exhibit their pro-tumorigenic effects by regulating immunosurveillance (immune suppression), differentiating into stromal cells (thereby contributing to the tumour microenvironment), promoting angiogenesis and stimulating an epithelial-mesenchymal transition, whereas inhibition of tumour growth by MSCs is reported to be through the inhibition of survival signalling pathways such as Akt and Wnt/β-catenin. The ability of MSCs to engraft and secrete cytokines at tumour sites has made them an attractive candidate to be engineered and used for delivery of antitumour agents. However, how tumour cells and MSCs crosstalk with each other is largely dependent on experimental factors as assessed in this review. Understanding these interactions through carefully designed experiments performed under controlled conditions which eliminate the variables alluded to above will help to understand the molecular basis of the effect of naïve MSCs on tumour growth. Furthermore, alternative strategies involving the modification of MSCs through genetic engineering with exogenous anticancer genes for the expression and/or secretion of a desired inhibitory factor could be exploited as a tool for developing a safer and more effective anticancer therapy.

Acknowledgements This research was funded by the South African Medical Research Council in terms of theSAMRC's Flagship Award Project SAMRC-RFA-UFSP-01-2013/STEM CELLS, the SAMRCExtramural Stem Cell Research and Therapy Unit, the National Research Foundation ofSouth Africa (grant no. 86942), the National Health Laboratory Services Research Trust (grant no. 94453), the University of Pretoria Research Development Programme (A0Z778), the University of Pretoria Vice Chancellor's Postdoctoral Fellowship and the Institute for Cellular and Molecular Medicine of the University of Pretoria.

Competing Interests No conflicts of interest, financial or otherwise, are declared by the authors.

Authors' Contribution MSP conceptualized the idea of the review, AKO and MAA did the literature search, AKO and MAA analysed the data, AKO and MAA prepared the manuscript, MSP edited and reviewed the drafted manuscript and AKO, MAA and MSP approved of the final version of the manuscript.

Supplementary Tables

Table S1a In vitro experiments that reported a stimulatory effect of BM-MSCs on breast cancer cells

Source of BM-MSCs	Type of experiment	Source of cancer cells	MSC status	References
Mouse	In vitro	Mouse	Naïve	Halpern et al. (2011)
Human	In vitro	Human	Naïve	Patel et al. (2010)
Human	In vitro	Human	Naïve	Sasser et al. (2007)
Mouse	In vitro	Mouse	Naïve	Zhang et al. (2013)
Human	In vitro	Human	Naïve	Hung et al. (2013)
Human	In vitro	Human	Naïve	De Luca et al. (2012)
Human	In vitro	Human	Naïve	Molloy et al. (2009)
Human	In vitro	Human	Naïve	Klopp et al. (2010)
Human	In vitro	Human	Naïve	Zhao et al. (2015)
Human	In vitro	Human	Naïve	Cuiffo et al. (2014)
Human	In vitro	Human	Engineered to produce TGFBR2	Shin et al. (2010)
Human	In vitro	Human	Naïve	Tobar et al. (2014)

Table S1b In vivo experiments that reported a stimulatory effect of BM-MSCs on breast cancer cells

Source of BM-MSCs	Type of experiment	Source of cancer cells	MSC status	References
Human	In vivo	Human	Naïve	Albarenque et al. (2011)
Human	In vivo	Human	Naïve	Rhodes et al. (2010)
Mouse	In vivo	Mouse	Naïve	Ke et al. (2013)
Human	In vivo	Human	Naïve	Cuiffo et al. (2014)
Mouse	In vivo	Mouse	Naïve	Yu et al. (2017)

Table S2a In vitro experiments reporting an inhibitory effect of BM-MSCs on breast cancer cells

Source of BM-MSCs	Type of experiment	Source of cancer cells	MSC status	References
Human	In vitro	Mouse	Naïve	Kéramidas et al. (2013)
Human	In vitro	Human	Naïve	Clarke et al. (2015)
Human	In vitro	Human	Naïve	Ono et al. (2014)
Mouse	In vitro	Mouse	Naïve	Lee et al. (2013)
Mouse	In vitro	Human	Naïve	Usha et al. (2013)
Human	In vitro	Human	Naïve	Lee et al. (2012)

Table S2b In vivo experiments reporting an inhibitory effect of BM-MSCs on breast cancer cells

Source of BM-MSCs	Type of experiment	Source of cancer cells	MSC status	References
Human	In vivo	Human	Engineered to produce BMP9	Wan et al. (2014)
Mouse	In vivo	Mouse	Naïve	Lee et al. (2013)
Mouse	In vivo	Human	Naïve	Usha et al. (2013)
Human	In vivo	Human	Naïve	Lee et al. (2012)

Table S3a In vitro experiments reporting a stimulatory effect of UC-MSCs on tumour growth

Source of UC-MSCs	Type of experiment	Source of cancer cells	Type of cancer cell	MSC status	References
Human	In vitro	Human	Oesophageal	Naïve	Yang et al. (2014a)
Human	In vitro	Human	Breast	Naïve	Di et al. (2014)
Human	In vitro	Human	Breast	Naïve	Ma et al. (2015)

Table S3b In vivo experiments reporting a stimulatory effect of UC-MSCs on tumour growth

Source of UC-MSCs	Type of experiment	Source of cancer cells	Type of cancer cell	MSC status	References
Human	In vivo	Human	Oesophageal	Naïve	Yang et al. (2014c)
Human	In vivo	Human	Gastric	[a] Engineered	Yang et al. (2014b)
Human	In vivo	Human	Breast	Naïve	Di et al. (2014)
Human	In vivo	Human	Breast	Naïve	Ma et al.(2015)
Human	In vivo	Mouse	Breast	Naïve	Yu et al. (2017)

[a]Engineered here refers to UC-MSC activated by macrophages to produce inflammatory cytokines

Table S4a In vitro experiments reporting an inhibitory effect of UC-MSCs on tumour growth

Source of UC-MSCs	Type of experiment	Source of cancer cells	Type of cancer cell	MSC status	References
Human	In vitro	Human	Breast	Naïve	Fong et al. (2011)
Human	In vitro	Human	Colorectal	Naïve	Fong et al. (2011)
Human	In vitro	Human	Hepatic	Naïve	Fong et al. (2011)
Rat	In vitro	Rat	Breast	Naïve	Kawabata et al. (2013)
Human	In vitro	Human	Bladder	Naïve	Wu et al. (2013)
Rat	In vitro	Mouse	Lung	Naïve	Maurya et al. (2010)
Rat	In vitro	Rat	Breast	Naïve	Ganta et al. (2009)
Human	In vitro	Human	Breast	Engineered to express IFN-β	Rachakatla et al. (2008)
Human	In vitro	Human	Glioma	Naïve	Yang et al. (2014a)
Human	In vitro	Human	Breast	Naïve	Chao et al. (2012)
Human	In vitro	Human	Myeloma	Naïve	Ciavarella et al. (2015)
Human	In vitro	Human	Prostate	Naïve	Han et al. (2014)

Table S4b In vivo experiments reporting an inhibitory effect of UC-MSCs on tumour growth

Source of UC-MSCs	Type of experiment	Source of cancer cells	Type of cancer cell	MSC status	References
Rat	In vivo	Rat	Breast	Naïve	Kawabata et al. (2013)
Human	In vivo	Human	Bladder	Naïve	Wu et al. (2013)
Rat	In vivo	Mouse	Lung	Naïve	Maurya et al. (2010)
Rat	In vivo	Rat	Breast	Naïve	Ganta et al. (2009)
Human	In vivo	Human	Breast	Engineered to express IFN-β	Rachakatla et al. (2008)
Human	In vivo	Human	Lung	Naïve	Rachakatla et al. (2007)
Human	In vivo	Human	Lung	Engineered to express human IFN-β	Rachakatla et al. (2007)
Human	In vivo	Human	Breast	Naïve	Chao et al. (2012)
Human	In vivo	Human	Myeloma	Naïve	Ciavarella et al. (2015)
Human	In vivo	Human	Prostate	Naïve	Han et al. (2014)

Table S5a In vitro experiments reporting a stimulatory effect of AD-MSC on tumour growth

Source of AD-MSCs	Type of experiment	Source of cancer cells	Type of cancer cell	MSC status	References
Human	In vitro	Human	Melanoma	Naïve	Kucerova et al. (2010)
Human	In vitro	Human	Glioma	Naïve	Kucerova et al. (2010)
Human	In vitro	Human	Glioma	Naïve	Yu et al. (2008)
Human	In vitro	Human	Lung	Naïve	Park et al. (2013)
Human	In vitro	Human	Breast	Naïve	Kamat et al. (2015)
Human	In vitro	Human	Head and neck	Naïve	Scherzed et al. (2013)
Human	In vitro	Human	Gastric	Naïve	Nomoto-Kojima et al. (2011)
Rat	In vitro	Human	Gastric	Naïve	Nomoto-Kojima et al. (2011)
Human	In vitro	Human	Breast	Naïve	Chen et al. (2014a)
Human	In vitro	Human	Breast	Naïve	Lin et al. (2013)
Human	In vitro	Human	Breast	Naïve	Zhao et al. (2012)
Human	In vitro	Human	Melanoma	Naïve	Kucerova et al. 2014)
Human	In vitro	Human	Sarcoma	Naïve	Bonuccelli et al. (2014)
Human	In vitro	Human	Breast	Naïve	Senst et al. (2013)
Human	In vitro	Human	Breast	Naïve	Xu et al. (2012)
Human	In vitro	Human	Ovarian	Naïve	Zhang et al. (2017)

Table S5b In vivo experiments reporting a stimulatory effect of AD-MSC on tumour growth

Source of AD-MSCs	Type of experiment	Source of cancer cells	Type of cancer cell	MSC status	References
Human	In vivo	Human	Melanoma	Naïve	Kucerova et al. (2010)
Human	In vivo	Human	Glioma	Naïve	Kucerova et al. (2010)
Human	In vivo	Human	Lung	Naïve	Yu et al. (2008)
Human	In vivo	Human	Glioma	Naïve	Yu et al. (2008)
Human	In vivo	Human	Melanoma	Naïve	Kucerova et al. (2014)
Human	In vivo	Human	Breast	Naïve	Yu et al. (2017)

Table S6a In vitro experiments reporting an inhibitory effect of AD-MSCs on tumour growth

Source of AD-MSCs	Type of experiment	Source of cancer cells	Type of cancer cell	MSC status	References
Human	In vitro	Human	Glioma	Naïve	Yang et al. (2014c)
Human	In vivo	Human	Melanoma	Engineered (CD-MSC/5FC)	Kucerova et al. (2014)
Human	In vitro	Human	Melanoma	Engineered (CD-MSC/5FC)	Kucerova et al. (2014)
Human	In vitro	Human	Hepatic	Naïve	Zhao et al. (2012)
Human	In vitro	Human	Lymphoma	Naïve	Ahn et al. (2014)
Human	In vitro	Human	Bladder	Naïve	Yu et al. (2016)
Human	In vitro	Human	Breast	Naïve	Zhao et al. (2013)
Human	In vitro	Human	Glioma	Engineered to secrete BMP4	Li et al. (2014)
Human	In vitro	Human	Glioma	Naïve	Li et al. (2014)
Human	In vitro	Human	Breast	Naïve	Yu et al. (2017)

CD-MSC/5FC represents MSC expressing fusion yeast cytosine deaminase::uracil phosphoribosyltransferase (CD-MSC) in combination with 5-fluorocytosine (5FC)

Table S6b In vivo experiments reporting an inhibitory effect of AD-MSCs on tumour growth

Source of AD-MSCs	Type of experiment	Source of cancer cells	Type of cancer cell	MSC status	References
Human	In vivo	Human	Melanoma	Engineered (CD-MSC/5FC)	Kucerova et al. (2014)
Mouse	In vivo	Mouse	Prostate	Engineered (CD-MSC)	Abrate et al. (2014)
Human	In vivo	Mouse	Prostate	Engineered (CD-MSC)	Abrate et al. (2014)
Human	In vivo	Human	Lymphoma	Naïve	Ahn et al. (2014)
Human	In vivo	Human	Glioma	Engineered to secrete BMP4	Li et al. (2014)
Human	In vivo	Human	Glioma	Naïve	Li et al. (2014)

CD-MSC/5FC represents MSC express fusion yeast cytosine deaminase/uracil phosphoribosyltransferase (CD-MSC) in combination with 5-fluorocytosine (5FC)

References

Abrate, A., Buono, R., Canu, T., Esposito, A., Del Maschio, A., Lucianò, R., Bettiga, A., Colciago, G., Guazzoni, G., Benigni, F., Hedlund, P., Altaner, C., Montorsi, F., & Cavarretta, I. T. R. (2014). Mesenchymal stem cells expressing therapeutic genes induce autochthonous prostate tumour regression. *European Journal of Cancer, 50*, 2478–2488.

Ahn, J. O., Chae, J. S., Coh, Y. R., Jung, W. S., Lee, H. W., Shin, I. S., Kang, S. K., & Youn, H. Y. (2014). Human adipose tissue-derived mesenchymal stem cells inhibit T-cell lymphoma growth in vitro and in vivo. *Anticancer Research, 34*, 4839–4847.

Albarenque, S. M., Zwacka, R. M., & Mohr, A. (2011). Both human and mouse mesenchymal stem cells promote breast cancer metastasis. *Stem Cell Research, 7*, 163–171.

Bianco, P., Cao, X., Frenette, P. S., Mao, J. J., Robey, P. G., Simmons, P. J., & Wang, C.-Y. (2013). The meaning, the sense and the significance: Translating the science of mesenchymal stem cells into medicine. *Nature Medicine, 19*, 35–42.

Bock, B. C., Stein, U., Schmitt, C. A., & Augustin, H. G. (2014). Mouse models of human cancer. *Cancer Research, 74*, 4671–4675.

Bonuccelli, G., Avnet, S., Grisendi, G., Salerno, M., Granchi, D., Dominici, M., Kusuzaki, K., & Baldini, N. (2014). Role of mesenchymal stem cells in osteosarcoma and metabolic reprogramming of tumor cells. *Oncotarget, 5*, 7575–7588.

Chao, K. C., Yang, H. T., & Chen, M. W. (2012). Human umbilical cord mesenchymal stem cells suppress breast cancer tumourigenesis through direct cell-cell contact and internalization. *Journal of Cellular and Molecular Medicine, 16*, 1803–1815.

Chen, Z., He, X., He, X., Chen, X., Lin, X., Zou, Y., Wu, X., & Lan, P. (2014a). Bone marrow mesenchymal stem cells ameliorate colitis-associated tumorigenesis in mice. *Biochemical and Biophysical Research Communications, 450*, 1402–1408.

Chen, D.-R., Lu, D.-Y., Lin, H.-Y., & Yeh, W.-L. (2014b). Mesenchymal stem cell-induced doxorubicin resistance in triple negative breast cancer. *BioMed Research International, 2014*, 10.

Ciavarella, S., Caselli, A., Tamma, A. V., Savonarola, A., Loverro, G., Paganelli, R., Tucci, M., & Silvestris, F. (2015). A peculiar molecular profile of umbilical cord-mesenchymal stromal cells drives their inhibitory effects on multiple myeloma cell growth and tumor progression. *Stem Cells and Development, 24*, 1457–1470.

Clarke, M. R., Imhoff, F. M., & Baird, S. K. (2015). Mesenchymal stem cells inhibit breast cancer cell migration and invasion through secretion of tissue inhibitor of metalloproteinase-1 and -2. *Molecular Carcinogenesis, 54*, 1214–1219.

Cuiffo, B. G., Campagne, A., Bell, G. W., Lembo, A., Orso, F., Lien, E. C., Bhasin, M. K., Raimo, M., Hanson, S. E., Marusyk, A., El-Ashry, D., Hematti, P., Polyak, K., Mechta-Grigoriou, F., Mariani, O., Volinia, S., Vincent-Salomon, A., Taverna, D., & Karnoub, A. E. (2014). MSC-regulated microRNAs converge on the transcription factor FOXP2 and promote breast cancer metastasis. *Cell Stem Cell, 15*, 762–774.

De Boeck, A., Pauwels, P., Hensen, K., Rummens, J. L., Westbroek, W., Hendrix, A., Maynard, D., Denys, H., Lambein, K., Braems, G., Gespach, C., Bracke, M., & De Wever, O. (2013). Bone marrow-derived mesenchymal stem cells promote colorectal cancer progression through paracrine neuregulin 1/HER3 signalling. *Gut, 62*, 550–560.

De Luca, A., Lamura, L., Gallo, M., Maffia, V., & Normanno, N. (2012). Mesenchymal stem cell-derived interleukin-6 and vascular endothelial growth factor promote breast cancer cell migration. *Journal of Cellular Biochemistry, 113*, 3363–3370.

Di, G. H., Liu, Y., Lu, Y., Liu, J., Wu, C., & Duan, H. F. (2014). IL-6 secreted from senescent mesenchymal stem cells promotes proliferation and migration of breast cancer cells. *PLoS One, 9*, e113572.

Dittmer, J., & Leyh, B. (2014). Paracrine effects of stem cells in wound healing and cancer progression. *International Journal of Oncology, 44*, 1789–1798.

Djouad, F., Plence, P., Bony, C., Tropel, P., Apparailly, F., Sany, J., Noel, D., & Jorgensen, C. (2003). Immunosuppressive effect of mesenchymal stem cells favors tumor growth in allogeneic animals. *Blood, 102*, 3837–3844.

Ferlay, J., Shin, H. R., Bray, F., Forman, D., Mathers, C., & Parkin, D. M. (2010). Estimates of worldwide burden of cancer in 2008: GLOBOCAN 2008. *International Journal of Cancer, 127*, 2893–2917.

Fong, C. Y., Chak, L. L., Biswas, A., Tan, J. H., Gauthaman, K., Chan, W. K., & Bongso, A. (2011). Human Wharton's jelly stem cells have unique transcriptome profiles compared to human embryonic stem cells and other mesenchymal stem cells. *Stem Cell Reviews, 7*, 1–16.

Ganta, C., Chiyo, D., Ayuzawa, R., Rachakatla, R., Pyle, M., Andrews, G., Weiss, M., Tamura, M., & Troyer, D. (2009). Rat umbilical cord stem cells completely abolish rat mammary carcinomas with no evidence of metastasis or recurrence 100 days post-tumor cell inoculation. *Cancer Research, 69*, 1815–1820.

Halpern, J. L., Kilbarger, A., & Lynch, C. C. (2011). Mesenchymal stem cells promote mammary cancer cell migration in vitro via the CXCR2 receptor. *Cancer Letters, 308*, 91–99.

Han, I., Yun, M., Kim, E. O., Kim, B., Jung, M. H., & Kim, S. H. (2014). Umbilical cord tissue-derived mesenchymal stem cells induce apoptosis in PC-3 prostate cancer cells through activation of JNK and downregulation of PI3K/AKT signaling. *Stem Cell Research & Therapy, 5*, 54.

Hong, I. S., Lee, H. Y., & Kang, K. S. (2014). Mesenchymal stem cells and cancer: Friends or enemies? *Mutation Research, 768*, 98–106.

Hung, S. P., Yang, M. H., Tseng, K. F., & Lee, O. K. (2013). Hypoxia-induced secretion of TGF-beta1 in mesenchymal stem cell promotes breast cancer cell progression. *Cell Transplantation, 22*, 1869–1882.

Kamat, P., Schweizer, R., Kaenel, P., Salemi, S., Calcagni, M., Giovanoli, P., Gorantla, V. S., Eberli, D., Andres, A. C., & Plock, J. A. (2015). Human adipose-derived Mesenchymal stromal cells may promote breast cancer progression and metastatic spread. *Plastic and Reconstructive Surgery, 136*, 76–84.

Kawabata, A., Ohta, N., Seiler, G., Pyle, M. M., Ishiguro, S., Zhang, Y. Q., Becker, K. G., Troyer, D., & Tamura, M. (2013). Naïve rat umbilical cord matrix stem cells significantly attenuate mammary tumor growth through modulation of endogenous immune responses. *Cytotherapy, 15*, 586–597.

Ke, C. C., Liu, R. S., Suetsugu, A., Kimura, H., Ho, J. H., Lee, O. K., & Hoffman, R. M. (2013). In vivo fluorescence imaging reveals the promotion of mammary tumorigenesis by mesenchymal stromal cells. *PLoS One, 8*, e69658.

Kéramidas, M., de Fraipont, F., Karageorgis, A., Moisan, A., Persoons, V., Richard, M.-J., Coll, J.-L., & Rome, C. (2013). The dual effect of mesenchymal stem cells on tumour growth and tumour angiogenesis. *Stem Cell Research & Therapy, 4*, 41–41.

Kidd, S., Spaeth, E., Dembinski, J. L., Dietrich, M., Watson, K., Klopp, A., Battula, V. L., Weil, M., Andreeff, M., & Marini, F. C. (2009). Direct evidence of mesenchymal stem cell tropism for tumor and wounding microenvironments using in vivo bioluminescent imaging. *Stem cells (Dayton, Ohio), 27*, 2614–2623.

Klopp, A. H., Lacerda, L., Gupta, A., Debeb, B. G., Solley, T., Li, L., Spaeth, E., Xu, W., Zhang, X., Lewis, M. T., Reuben, J. M., Krishnamurthy, S., Ferrari, M., Gaspar, R., Buchholz, T. A., Cristofanilli, M., Marini, F., Andreeff, M., & Woodward, W. A. (2010). Mesenchymal stem cells promote mammosphere formation and decrease E-cadherin in normal and malignant breast cells. *PLoS One, 5*, e12180.

Kucerova, L., Matuskova, M., Hlubinova, K., Altanerova, V., & Altaner, C. (2010). Tumor cell behaviour modulation by mesenchymal stromal cells. *Molecular Cancer, 9*, 129.

Kucerova, L., Skolekova, S., Demkova, L., Bohovic, R., & Matuskova, M. (2014). Long-term efficiency of mesenchymal stromal cell-mediated CD-MSC/5FC therapy in human melanoma xenograft model. *Gene Therapy, 21*, 874–887.

Lazennec, G., & Lam, P. Y. (2016). Recent discoveries concerning the tumor – mesenchymal stem cell interactions. *Biochimica et Biophysica Acta, 1866*, 290–299.

Lee, R. H., Yoon, N., Reneau, J. C., & Prockop, D. J. (2012). Preactivation of human MSCs with TNF-alpha enhances tumor-suppressive activity. *Cell Stem Cell, 11*, 825–835.

Lee, M. J., Heo, S. C., Shin, S. H., Kwon, Y. W., Do, E. K., Suh, D. S., Yoon, M. S., & Kim, J. H. (2013). Oncostatin M promotes mesenchymal stem cell-stimulated tumor growth through a paracrine mechanism involving periostin and TGFBI. *The International Journal of Biochemistry & Cell Biology, 45*, 1869–1877.

Li, Q., Wijesekera, O., Salas, S. J., Wang, J. Y., Zhu, M., Aprhys, C., Chaichana, K. L., Chesler, D. A., Zhang, H., Smith, C. L., Guerrero-Cazares, H., Levchenko, A., & Quinones-Hinojosa, A. (2014). Mesenchymal stem cells from human fat engineered to secrete BMP4 are nononcogenic, suppress brain cancer, and prolong survival. *Clinical Cancer Research: An Official Journal of the American Association for Cancer Research, 20*, 2375–2387.

Lin, R., Wang, S., & Zhao, R. C. (2013). Exosomes from human adipose-derived mesenchymal stem cells promote migration through Wnt signaling pathway in a breast cancer cell model. *Molecular and Cellular Biochemistry, 383*, 13–20.

Ljujic, B., Milovanovic, M., Volarevic, V., Murray, B., Bugarski, D., Przyborski, S., Arsenijevic, N., Lukic, M. L., & Stojkovic, M. (2013). Human mesenchymal stem cells creating an immunosuppressive environment and promote breast cancer in mice. *Scientific Reports, 3*, 2298.

Loebinger, M. R., Kyrtatos, P. G., Turmaine, M., Price, A. N., Pankhurst, Q., Lythgoe, M. F., & Janes, S. M. (2009). Magnetic resonance imaging of mesenchymal stem cells homing to pulmonary metastases using biocompatible magnetic nanoparticles. *Cancer Research, 69*, 8862–8867.

Lopez, M. J., & Spencer, N. D. (2011). In vitro adult rat adipose tissue-derived stromal cell isolation and differentiation. *Methods in Molecular Biology, 702*, 37–46.

Ma, F., Chen, D., Chen, F., Chi, Y., Han, Z., Feng, X., Li, X., & Han, Z. (2015). Human umbilical cord Mesenchymal stem cells promote breast cancer metastasis by Interleukin-8- and Interleukin-6-dependent induction of CD44 (+)/CD24(−) cells. *Cell Transplantation, 24*, 2585–2599.

Maurya, D. K., Doi, C., Kawabata, A., Pyle, M. M., King, C., Wu, Z., Troyer, D., & Tamura, M. (2010). Therapy with un-engineered naive rat umbilical cord matrix stem cells markedly inhibits growth of murine lung adenocarcinoma. *BMC Cancer, 10*, 590.

Menon, L. G., Picinich, S., Koneru, R., Gao, H., Lin, S. Y., Koneru, M., Mayer-Kuckuk, P., Glod, J., & Banerjee, D. (2007). Differential gene expression associated with migration of mesenchymal stem cells to conditioned medium from tumor cells or bone marrow cells. *Stem Cells (Dayton, Ohio), 25*, 520–528.

Molloy, A. P., Martin, F. T., Dwyer, R. M., Griffin, T. P., Murphy, M., Barry, F. P., O'Brien, T., & Kerin, M. J. (2009). Mesenchymal stem cell secretion of chemokines during differentiation into osteoblasts, and their potential role in mediating interactions with breast cancer cells. *International Journal of Cancer, 124*, 326–332.

Motaln, H., Schichor, C., & Lah, T. T. (2010). Human mesenchymal stem cells and their use in cell-based therapies. *Cancer, 116*, 2519–2530.

Nakamizo, A., Marini, F., Amano, T., Khan, A., Studeny, M., Gumin, J., Chen, J., Hentschel, S., Vecil, G., Dembinski, J., Andreeff, M., & Lang, F. F. (2005). Human bone marrow-derived mesenchymal stem cells in the treatment of gliomas. *Cancer Research, 65*, 3307–3318.

Nakamura, K., Ito, Y., Kawano, Y., Kurozumi, K., Kobune, M., Tsuda, H., Bizen, A., Honmou, O., Niitsu, Y., & Hamada, H. (2004). Antitumor effect of genetically engineered mesenchymal stem cells in a rat glioma model. *Gene Therapy, 11*, 1155–1164.

Nomoto-Kojima, N., Aoki, S., Uchihashi, K., Matsunobu, A., Koike, E., Ootani, A., Yonemitsu, N., Fujimoto, K., & Toda, S. (2011). Interaction between adipose tissue stromal cells and gastric cancer cells in vitro. *Cell and Tissue Research, 344*, 287–298.

Ono, M., Kosaka, N., Tominaga, N., Yoshioka, Y., Takeshita, F., Takahashi, R. U., Yoshida, M., Tsuda, H., Tamura, K., & Ochiya, T. (2014). Exosomes from bone marrow mesenchymal stem cells contain a microRNA that promotes dormancy in metastatic breast cancer cells. *Science Signaling, 7*, ra63.

Paris, J. L., de la Torre, P., Manzano, M., Cabanas, M. V., Flores, A. I., & Vallet-Regi, M. (2016). Decidua-derived mesenchymal stem cells as carriers of mesoporous silica nanoparticles. In vitro and in vivo evaluation on mammary tumors. *Acta Biomaterialia, 33*, 275–282.

Park, Y. M., Yoo, S. H., & Kim, S. H. (2013). Adipose-derived stem cells induced EMT-like changes in H358 lung cancer cells. *Anticancer Research, 33*, 4421–4430.

Patel, S. A., Meyer, J. R., Greco, S. J., Corcoran, K. E., Bryan, M., & Rameshwar, P. (2010). Mesenchymal stem cells protect breast cancer cells through regulatory T cells: Role of mesenchymal stem cell-derived TGF-beta. *Journal of immunology (Baltimore, Md: 1950), 184*, 5885–5894.

Quante, M., Tu, S. P., Tomita, H., Gonda, T., Wang, S. S., Takashi, S., Baik, G. H., Shibata, W., Diprete, B., Betz, K. S., Friedman, R., Varro, A., Tycko, B., & Wang, T. C. (2011). Bone marrow-derived myofibroblasts contribute to the mesenchymal stem cell niche and promote tumor growth. *Cancer Cell, 19*, 257–272.

Rachakatla, R. S., Marini, F., Weiss, M. L., Tamura, M., & Troyer, D. (2007). Development of human umbilical cord matrix stem cell-based gene therapy for experimental lung tumors. *Cancer Gene Therapy, 14*, 828–835.

Rachakatla, R. S., Pyle, M. M., Ayuzawa, R., Edwards, S. M., Marini, F. C., Weiss, M. L., Tamura, M., & Troyer, D. (2008). Combination treatment of human umbilical cord matrix stem cell-based interferon-beta gene therapy and 5-fluorouracil significantly reduces growth of metastatic human breast cancer in SCID mouse lungs. *Cancer Investigation, 26*, 662–670.

Revell, C. M., & Athanasiou, K. A. (2009). Success rates and immunologic responses of autogenic, allogenic, and xenogenic treatments to repair articular cartilage defects. *Tissue Engineering, Part B: Reviews, 15*, 1–15.

Rhodes, L. V., Antoon, J. W., Muir, S. E., Elliott, S., Beckman, B. S., & Burow, M. E. (2010). Effects of human mesenchymal stem cells on ER-positive human breast carcinoma cells mediated through ER-SDF-1/CXCR4 crosstalk. *Molecular Cancer, 9*, 295.

Ridge, S. M., Sullivan, F. J., & Glynn, S. A. (2017). Mesenchymal stem cells: Key players in cancer progression. *Molecular Cancer, 16*, 31.

Sasportas, L. S., Kasmieh, R., Wakimoto, H., Hingtgen, S., van de Water, J. A., Mohapatra, G., Figueiredo, J. L., Martuza, R. L., Weissleder, R., & Shah, K. (2009). Assessment of therapeutic efficacy and fate of engineered human mesenchymal stem cells for cancer therapy. *Proceedings of the National Academy of Sciences of the United States of America, 106*, 4822–4827.

Sasser, A. K., Mundy, B. L., Smith, K. M., Studebaker, A. W., Axel, A. E., Haidet, A. M., Fernandez, S. A., & Hall, B. M. (2007). Human bone marrow stromal cells enhance breast cancer cell growth rates in a cell line-dependent manner when evaluated in 3D tumor environments. *Cancer Letters, 254*, 255–264.

Scherzed, A., Hackenberg, S., Radeloff, A., Froelich, K., Rak, K., Hagen, R., & Kleinsasser, N. (2013). Human mesenchymal stem cells promote cancer motility and cytokine secretion in vitro. *Cells, Tissues, Organs, 198*, 327–337.

Senst, C., Nazari-Shafti, T., Kruger, S., Honer Zu Bentrup, K., Dupin, C. L., Chaffin, A. E., Srivastav, S. K., Worner, P. M., Abdel-Mageed, A. B., Alt, E. U., & Izadpanah, R. (2013). Prospective dual role of mesenchymal stem cells in breast tumor microenvironment. *Breast Cancer Research and Treatment, 137*, 69–79.

Serakinci, N., Fahrioglu, U., & Christensen, R. (2014). Mesenchymal stem cells, cancer challenges and new directions. *European Journal of Cancer, 50*, 1522–1530.

Shin, S. Y., Nam, J. S., Lim, Y., & Lee, Y. H. (2010). TNFalpha-exposed bone marrow-derived mesenchymal stem cells promote locomotion of MDA-MB-231 breast cancer cells through transcriptional activation of CXCR3 ligand chemokines. *The Journal of Biological Chemistry, 285*, 30731–30740.

Sigrist, S., Ebel, N., Langlois, A., Bosco, D., Toso, C., Kleiss, C., Mandes, K., Berney, T., Pinget, M., Belcourt, A., & Kessler, L. (2005). Role of chemokine signaling pathways in pancreatic islet rejection during allo- and xenotransplantation. *Transplantation Proceedings, 37*, 3516–3518.

Soerjomataram, I., Lortet-Tieulent, J., Parkin, D. M., Ferlay, J., Mathers, C., Forman, D., & Bray, F. (2012). Global burden of cancer in 2008: A systematic analysis of disability-adjusted life-years in 12 world regions. *Lancet, 380*, 1840–1850.

Squillaro, T., Peluso, G., & Galderisi, U. (2016). Clinical trials with Mesenchymal stem cells: An update. *Cell Transplantation, 25*, 829–848.

Studeny, M., Marini, F. C., Champlin, R. E., Zompetta, C., Fidler, I. J., & Andreeff, M. (2002). Bone marrow-derived mesenchymal stem cells as vehicles for interferon-beta delivery into tumors. *Cancer Research, 62*, 3603–3608.

Tholpady, S. S., Katz, A. J., & Ogle, R. C. (2003). Mesenchymal stem cells from rat visceral fat exhibit multipotential differentiation in vitro. *The Anatomical Record Part A: Discoveries in Molecular, Cellular, and Evolutionary Biology, 272*, 398–402.

Tobar, N., Avalos, M. C., Mendez, N., Smith, P. C., Bernabeu, C., Quintanilla, M., & Martinez, J. (2014). Soluble MMP-14 produced by bone marrow-derived stromal cells sheds epithelial endoglin modulating the migratory properties of human breast cancer cells. *Carcinogenesis, 35*, 1770–1779.

Usha, L., Rao, G., Christopherson Ii, K., & Xu, X. (2013). Mesenchymal stem cells develop tumor tropism but do not accelerate breast cancer tumorigenesis in a somatic mouse breast cancer model. *PLoS One, 8*, e67895.

Wagner, W., Wein, F., Seckinger, A., Frankhauser, M., Wirkner, U., Krause, U., Blake, J., Schwager, C., Eckstein, V., Ansorge, W., & Ho, A. D. (2005). Comparative characteristics of mesenchymal stem cells from human bone marrow, adipose tissue, and umbilical cord blood. *Experimental Hematology, 33*, 1402–1416.

Wan, S., Liu, Y., Weng, Y., Wang, W., Ren, W., Fei, C., Chen, Y., Zhang, Z., Wang, T., Wang, J., Jiang, Y., Zhou, L., He, T., & Zhang, Y. (2014). BMP9 regulates cross-talk between breast cancer cells and bone marrow-derived mesenchymal stem cells. *Cellular Oncology (Dordrecht), 37*, 363–375.

Wu, S., Ju, G. Q., Du, T., Zhu, Y. J., & Liu, G. H. (2013). Microvesicles derived from human umbilical cord Wharton's jelly mesenchymal stem cells attenuate bladder tumor cell growth in vitro and in vivo. *PLoS One, 8*, e61366.

Xu, Q., Wang, L., Li, H., Han, Q., Li, J., Qu, X., Huang, S., & Zhao, R. C. (2012). Mesenchymal stem cells play a potential role in regulating the establishment and maintenance of epithelial-mesenchymal transition in MCF7 human breast cancer cells by paracrine and induced autocrine TGF-beta. *International Journal of Oncology, 41*, 959–968.

Yang, C., Lei, D., Ouyang, W., Ren, J., Li, H., Hu, J., & Huang, S. (2014a). Conditioned media from human adipose tissue-derived Mesenchymal stem cells and umbilical cord-derived Mesenchymal stem cells efficiently induced the apoptosis and differentiation in human Glioma cell lines in vitro. *BioMed Research International, 2014*, 13.

Yang, X., Li, Z., Ma, Y., Gao, J., Liu, S., Gao, Y., & Wang, G. (2014b). Human umbilical cord mesenchymal stem cells promote carcinoma growth and lymph node metastasis when co-injected with esophageal carcinoma cells in nude mice. *Cancer Cell International, 14*, 93.

Yang, T., Zhang, X., Wang, M., Zhang, J., Huang, F., Cai, J., Zhang, Q., Mao, F., Zhu, W., Qian, H., & Xu, W. (2014c). Activation of mesenchymal stem cells by macrophages prompts human gastric cancer growth through NF-kappaB pathway. *PLoS One, 9*, e97569.

Yu, J. M., Jun, E. S., Bae, Y. C., & Jung, J. S. (2008). Mesenchymal stem cells derived from human adipose tissues favor tumor cell growth in vivo. *Stem Cells and Development, 17*, 463–473.

Yu, X., Su, B., Ge, P., Wang, Z., Li, S., Huang, B., Gong, Y., & Lin, J. (2016). AB237. Human adipose derived stem cells induced cell apoptosis and S phase arrest in bladder tumor. *Translational Andrology and Urology, 5*, AB237.

Yu, P. F., Huang, Y., Xu, C. L., Lin, L. Y., Han, Y. Y., Sun, W. H., Hu, G. H., Rabson, A. B., Wang, Y., & Shi, Y. F. (2017). Downregulation of CXCL12 in mesenchymal stromal cells by TGFbeta promotes breast cancer metastasis. *Oncogene, 36*, 840–849.

Zhang, T., Lee, Y. W., Rui, Y. F., Cheng, T. Y., Jiang, X. H., & Li, G. (2013). Bone marrow-derived mesenchymal stem cells promote growth and angiogenesis of breast and prostate tumors. *Stem Cell Research & Therapy, 4*, 70.

Zhang, Y., Dong, W., Wang, J., Cai, J., & Wang, Z. (2017). Human omental adipose-derived mesenchymal stem cell-conditioned medium alters the proteomic profile of epithelial ovarian cancer cell lines in vitro. *Oncotargets and Therapy, 10*, 1655–1663.

Zhao, M., Sachs, P. C., Wang, X., Dumur, C. I., Idowu, M. O., Robila, V., Francis, M. P., Ware, J., Beckman, M., Rizki, A., Holt, S. E., & Elmore, L. W. (2012). Mesenchymal stem cells in mammary adipose tissue stimulate progression of breast cancer resembling the basal-type. *Cancer Biology & Therapy, 13*, 782–792.

Zhao, Y., Gao, J., & Lu, F. (2013). Human adipose-derived stem cell adipogenesis induces paracrine regulation of the invasive ability of MCF-7 human breast cancer cells in vitro. *Experimental and Therapeutic Medicine, 6*, 937–942.

Zhao, Q., Gregory, C. A., Lee, R. H., Reger, R. L., Qin, L., Hai, B., Park, M. S., Yoon, N., Clough, B., McNeill, E., Prockop, D. J., & Liu, F. (2015). MSCs derived from iPSCs with a modified protocol are tumor-tropic but have much less potential to promote tumors than bone marrow MSCs. *Proceedings of the National Academy of Sciences of the United States of America, 112*, 530–535.

Zhou, Z., Chen, Y., Zhang, H., Min, S., Yu, B., He, B., & Jin, A. (2013). Comparison of mesenchymal stromal cells from human bone marrow and adipose tissue for the treatment of spinal cord injury. *Cytotherapy, 15*, 434–448.

Zischek, C., Niess, H., Ischenko, I., Conrad, C., Huss, R., Jauch, K. W., Nelson, P. J., & Bruns, C. (2009). Targeting tumor stroma using engineered mesenchymal stem cells reduces the growth of pancreatic carcinoma. *Annals of Surgery, 250*, 747–753.

Advs Exp. Medicine, Biology - Innovations in Cancer Research and Regenerative Medicine (2018) 1: 125–144
DOI 10.1007/5584_2017_119

Published online: 15 November 2017

The Role of Reactive Oxygen Species in Adipogenic Differentiation

Danielle de Villiers, Marnie Potgieter, Melvin A. Ambele, Ladislaus Adam, Chrisna Durandt, and Michael S. Pepper

Abstract

Interest in reactive oxygen species and adipocyte differentiation/adipose tissue function is steadily increasing. This is due in part to a search for alternative avenues for combating obesity, which results from the excess accumulation of adipose tissue. Obesity is a major risk factor for complex disorders such as cancer, type 2 diabetes, and cardiovascular diseases. The ability of mesenchymal stromal/stem cells (MSCs) to differentiate into adipocytes is often used as a model for studying adipogenesis in vitro. A key focus is the effect of both intra- and extracellular reactive oxygen species (ROS) on adipogenesis. The consensus from the majority of studies is that ROS, irrespective of the source, promote adipogenesis.

The effect of ROS on adipogenesis is suppressed by antioxidants or ROS scavengers. Reactive oxygen species are generated during the process of adipocyte differentiation as well as by other cell metabolic processes. Despite many studies in this field, it is still not possible to state with certainty whether ROS measured during adipocyte differentiation are

D. de Villiers, L. Adam, C. Durandt, and M.S. Pepper (✉)
Department of Immunology and Institute for Cellular and Molecular Medicine; SAMRC Extramural Unit for Stem Cell Research and Therapy; Faculty of Health Sciences, University of Pretoria, Pretoria, South Africa
e-mail: danielledevilliers@gmail.com; tapsltal@yahoo.com; chrisna.durandt@up.ac.za; michael.pepper@up.ac.za

M. Potgieter
Department of Immunology and Institute for Cellular and Molecular Medicine; SAMRC Extramural Unit for Stem Cell Research and Therapy; Faculty of Health Sciences, University of Pretoria, Pretoria, South Africa

Center for Microbial Ecology and Genomics, Department of Genetics, Faculty of Natural and Agricultural Sciences, University of Pretoria, Pretoria, South Africa
e-mail: marniepotgieter@yahoo.com

M.A. Ambele
Department of Immunology and Institute for Cellular and Molecular Medicine; SAMRC Extramural Unit for Stem Cell Research and Therapy; Faculty of Health Sciences, University of Pretoria, Pretoria, South Africa

Department of Oral Pathology and Oral Biology, School of Dentistry, Faculty of Health Sciences, University of Pretoria, Pretoria, South Africa
e-mail: ambelema@gmail.com

a cause or consequence of this process. In addition, it is still unclear what the exact sources are of the ROS that initiate and/or drive adipogenic differentiation in MSCs in vivo. This review provides an overview of our understanding of the role of ROS in adipocyte differentiation as well as how certain ROS scavengers and antioxidants might affect this process.

Keywords
Adipogenic differentiation • Adipose-derived stromal cells • Mesenchymal stem/stromal cells • Reactive oxygen species • ROS scavengers

Abbreviations

ASCs	Adipose-derived stem/stromal cells
ATP	Adenosine triphosphate
BAT	Brown adipose tissue
BMAL1	Brain and muscle ARNT-like protein 1
BM-MSCs	Bone marrow-derived MSCs
BMP	Bone morphogenic protein
C/EBP	CCAAT enhancer-binding protein
C/EBPα	CCAAT enhancer-binding protein alpha
C/EBPβ	CCAAT enhancer-binding protein beta
C/EBPδ	CCAAT enhancer-binding protein delta
CAT	Catalase
CCL2	C-C motif chemokine 2 precursor
CoQ	Oxidized ubiquinone
$CoQH_2$	Reduced ubiquinol
CREB	Cyclic AMP response element-binding protein
Cyt	Cytochrome
DPI	Diphenyleneiodonium
e^-	electron
eNOS	Endothelial nitric oxide
EPAS1	Endothelial PAS domain protein 1
ETC	Electron transport chain
FABP4	Fatty acid-binding protein 4
FAT	Fatty acid translocase (CD36)
FOXA2	Forkhead box A2
FOXO1	Forkhead box O1
Ga	One billion years
GATA2	GATA binding protein 2
GATA3	GATA binding protein 3
GPx	Glutathione peroxidase
GSTA4	Glutathione S-transferase A4
H^+	Proton
H_2O	Water
H_2O_2	Hydrogen peroxide
IDII	Standard adipogenic induction cocktail
IL10	Interleukin 10
IL6	Interleukin 6
IL8	Interleukin 8
iNOS	Inducible nitric oxide synthase
KLF	Kruppel-like factor
LPL	Lipoprotein lipase
M1	Classically activated macrophage phenotype
MEFs	Immortalized murine embryonic fibroblasts
mESCs	Murine embryonic stem cells
MKP-1	MAP kinase phosphatase-1
mMSCs	Murine mesenchymal stem/stromal cells
MSCs	Mesenchymal stem/stromal cells
NAC	N-acetyl-L-cysteine
NEFA	Nonesterified fatty acid
NO	Nitric oxide
NOS	Nitric oxide synthase
NOX	NADPH oxidase
O_2	Oxygen
O_2^-	Superoxide
PPARγ	Proliferator-activated receptor-gamma
Pref-1	Preadipocyte factor-1
Prx3	Peroxiredoxin 3
ROS	Reactive oxygen species
SIRT1	Histone deacetylase sirtuin 1
SOD	Superoxide dismutase
SOD2	Superoxide dismutase 2

SREBP1c	Sterol regulatory element-binding transcription factor 1
STAT5a	Signal transducer and activator of transcription-5a
TAZ	Transcriptional coactivator with PDZ-binding motif
TG	Triacylglycerol
TNFα	Tumor necrosis factor alpha
WAT	White adipose tissue
ZFP423	Zinc finger protein 423

1 Introduction

Adipogenesis can be described as the differentiation of stem cells to form fully differentiated lipid-filled mature adipocytes (Lefterova and Lazar 2009). It is a complex and well-orchestrated process that occurs through different stages involving numerous transcription factors, cell-cycle proteins, extracellular signals, hormones, and small molecules which mediate the up- and downregulation of a well-defined cascade of genes essential for the differentiation and maturation of adipocytes (Ali et al. 2013; Jiang et al. 2012).

Adipogenic differentiation has primarily been studied in vitro using murine preadipocyte cell lines such as 3T3-L1, but some studies have also used primary or immortalized murine embryonic fibroblasts (MEFs), murine embryonic stem cells (mESCs), and murine mesenchymal stem/stromal cells (mMSCs). A disadvantage of using murine cell lines is that the translation of the findings to the human setting is uncertain and challenging. As an example, the same repertoire of gene expression that occurs during human adipogenesis may not necessarily be present when murine cells are used. Another key difference between mouse and human adipose tissue is that the distribution of white and brown fat differs between the two species. Heterogeneity in the cellular composition of adipose tissue and inter-patient variability are some of the other main differences observed between humans and murine experimental models (Ambele et al. 2016). A human primary cell model would therefore provide better insight into adipose differentiation and metabolism, thus increasing our understanding of obesity and its related comorbidities in humans.

Reactive oxygen species (ROS) are known to play a role in promoting adipogenic differentiation of mouse, rat, and human preadipocytes as well as immortalized preadipocyte cell lines (Tormos et al. 2011; Kanda et al. 2011; Schroder et al. 2009; Lee et al. 2009). Excess accumulation of ROS in cells, as a result of limited scavenging activity by antioxidant systems, induces oxidative stress, which may lead to cellular damage of proteins, lipids, and nucleic acids (Atashi et al. 2015). This review will discuss the effect of ROS on adipogenesis using immortalized cell lines and primary cells of both human and murine origin.

2 Brief History of Redox Biology and ROS Involvement in Biological Processes

The accumulation of atmospheric oxygen (O_2) began around 2.3 billion (Ga) years ago (Bekker et al. 2004). The rise in atmospheric O_2 levels and the need for O_2 in aerobic life forms for cellular respiration are the fundamental tenets underlying redox biology. Oxygen can be converted through electron-transferring chemical reactions or exposure to environmental stresses such as ultraviolet radiation and heat, into a variety of highly reactive chemical forms that are collectively known as reactive oxygen species (ROS). Hydrogen peroxide (H_2O_2) was the first ROS discovered by Thénard in 1818, but the production of biological ROS was only described in 1954 (Commoner et al. 1954). Initially, ROS were simply considered to be by-products of aerobic metabolism that were toxic in high levels to biological systems through oxidation and nitration of macromolecules such as proteins, lipids, and nucleic acids. A few years later these damaging properties of ROS would be associated with aging (Harman 1956).

Subsequently, it was shown that the generation of ROS does not only have negative consequences, but constitutes an important intracellular signaling system that plays an essential role in biological processes such as differentiation, proliferation, cell death, and senescence (Liu et al. 2012a; D'Autreaux and Toledano 2007; Atashi et al. 2015; Kawagishi and Finkel 2014). An example is phagocytosis that requires a dramatic increase in O_2 uptake due to the demand for metabolic energy through glycolysis during the phagocytic process (Sbarra and Karnovsky 1959). Other immune cells, such as neutrophils and macrophages, have also been shown to produce high levels of ROS through an oxidative burst that contributes to innate host defense (Chen and Junger 2012; Slauch 2011).

Initially the mechanisms involved in ROS production were largely unknown. In 1967, Hinkle and colleagues discovered that the electron transport chain (ETC) of mitochondria produces H_2O_2 (Hinkle et al. 1967). This discovery was followed by the discovery in 1969 of the superoxide dismutase (SOD) enzyme and its ability to convert (dismutate) chemically generated superoxide (O_2^-) into O_2 and H_2O_2 (McCord and Fridovich 1969). In 1974, Loschen and colleagues discovered that O_2^- is a reactive precursor of mitochondrial H_2O_2 (Loschen et al. 1974). In 1961, Iyer and colleagues suggested that the phagocyte respiratory burst results in the generation of H_2O_2, but the mechanism involved was unknown (Bedard and Krause 2007; Jiang et al. 2011; Iyer et al. 1961). In 1964, Rossi and Zatti discovered that an NADPH oxidase (NOX) was responsible for the respiratory burst in phagocytes (Rossi and Zatti 1964). A few years later, in 1973, Babior and colleagues showed that the initial product of the respiratory burst is O_2^- and not H_2O_2 as had originally been proposed (Babior et al. 1973).

In 1991, Meier and colleagues discovered NOX enzyme systems in human fibroblasts (Meier et al. 1991). The discovery of these systems in non-phagocytic cells was a pivotal event in redox biology, as ROS now appeared to have functions other than just in host defense. Subsequently, gp91^{phox}, the catalytic subunit of the phagocyte NOX, and various other components of the NOX enzyme complex, including p22^{phox}, p40^{phox}, p47^{phox}, p67^{phox}, and others, were characterized (Bedard and Krause 2007; Jiang et al. 2011). The most important historical events in the discovery of ROS are summarized in Fig. 1.

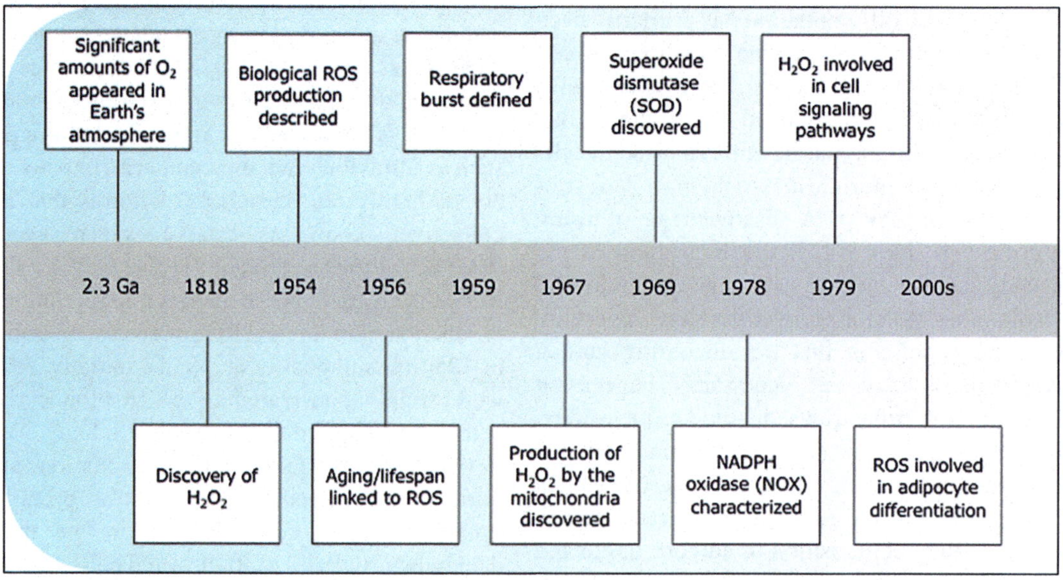

Fig. 1 A summary of the history of redox biology

3 Sources of ROS and their Mechanism of Production

The production of cellular ROS originates from many different sources including the mitochondrial ETC, NOX, xanthine oxidase, cytochrome p450, uncoupled nitric oxide synthase (NOS), myeloperoxidase, and others. However, the mitochondrial ETC and NOX enzymes remain the major sources of cellular ROS production.

3.1 Mitochondria

Mitochondria play an important role in the cellular metabolism of aerobes by facilitating the conversion of O_2 into adenosine triphosphate (ATP) and water (H_2O) (Holzerova and Prokisch 2015). Low levels of O_2^- are naturally produced via the ETC during this process (Holzerova and Prokisch 2015). Electron transport chain sites that are involved in ROS generation include Complexes I (NADH/ubiquinone oxidoreductase), II (succinate dehydrogenase), and III (cytochrome *bc1* complex) (Tahara et al. 2009; Sabharwal and Schumacker 2014). Complexes I and III are generally regarded as the main sites of O_2^- production in mitochondria (Drose and Brandt 2008). Although implicated in O_2^- generation, Complex II seems to be involved to a lesser extent when compared to Complexes I and III (Ryu et al. 2015). However, other studies suggest that Complex II is able to generate O_2^- at similar levels to those produced by Complex I and III (Quinlan et al. 2012). Complex IV accepts an electron from cytochrome *c* (Cyt c) and passes it to O_2, the final electron acceptor in this chain, to form H_2O. Water is formed by the reduction of O_2 with a series of four electrons, consecutively. Energy is released by electron transfer, which is used to pump protons (H^+) out of the matrix, through Complex I, III, and IV into the intermembrane space, building up a significant H^+ concentration gradient. The H^+ gradient that is generated provides the energy needed to produce ATP by ATP synthase (Complex V). Leakage of electrons to O_2 forming the one-electron reduction of O_2 to O_2^- and thus, its derivative ROS, occurs mainly at Complex I and III (Fig. 2). Superoxide is inherently unstable and is, soon after being produced, dismutated into the more stable H_2O_2. The dismutation of O_2^- can occur spontaneously or through the action of antioxidant enzymes such as Cu-, Mn-, and ZnSOD found in the mitochondrial intermembrane space (Turrens 2003; Finkel 2011). Hydrogen peroxide is then converted to H_2O and O_2 by glutathione peroxidase (GPx). Hydrogen peroxide can be scavenged by antioxidants such as catalase (CAT), GPx, and peroxiredoxin 3 (Prx3) or act as a signaling molecule in different signaling pathways including redox balance, energy metabolism, cell cycle, and the stress response (Atashi et al. 2015; Droge 2002; Finkel 2011). The production of ROS is summarized in Fig. 3.

Although the ETC has been identified as the main source of mitochondrial ROS generation, various other enzymes located within mitochondria play a role in ROS production. These enzymes include monoamine oxidase, α-ketoglutarate dehydrogenase, cytochrome b5 reductase, glycerol-3-phosphate dehydrogenase, various P450 enzymes, aconitase, pyruvate dehydrogenase, and others (Finkel 2011).

3.2 NADPH Oxidase

The NOX family is a group of NADPH oxidases that are responsible for transferring electrons, generating O_2^- in the process (Bedard and Krause 2007; Panday et al. 2015) (Fig. 4). This dedicated function of NOX differentiates this group of enzymes from other oxidoreductases, such as cyclooxygenase, lipoxygenase, cytochrome P450 enzymes, NOS, xanthine oxidase, mitochondrial NADPH/ubiquinone oxidoreductase (Complex I), and others, which produce ROS only as by-products during their catalytic action in their various pathways (Jiang et al. 2011).

The membrane protein, cytochrome b558, consists of two subunits, a large subunit $gp91^{phox}$ (NOX2) and a smaller subunit $p22^{phox}$, and forms the backbone of NOX. During cell activation, two regulatory subunits in the cytoplasm, $p47^{phox}$ and

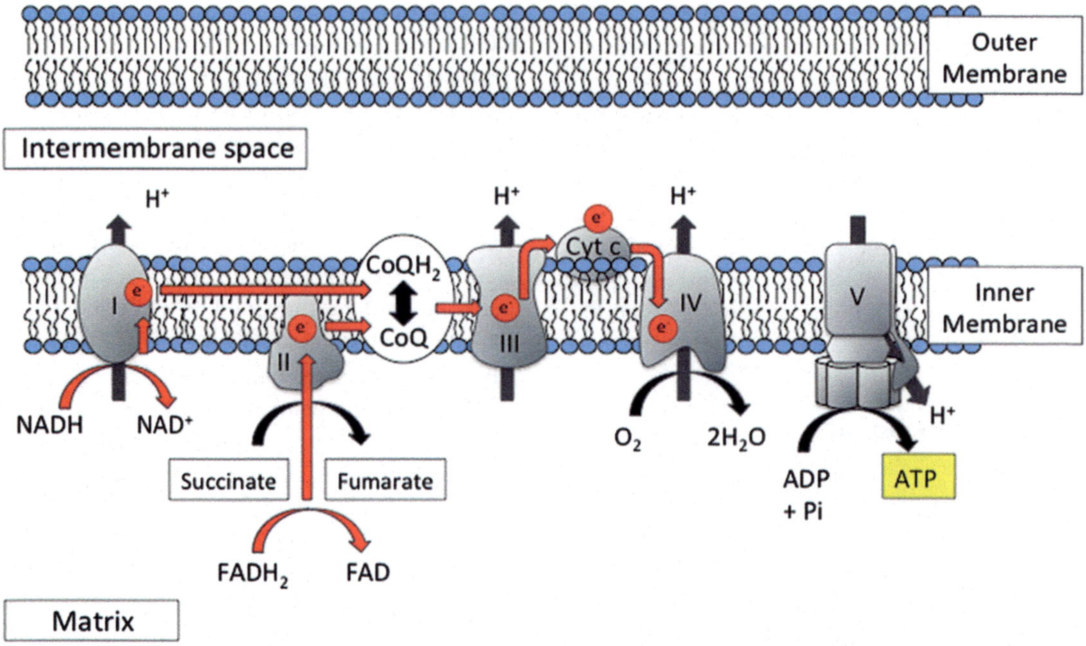

Fig. 2 The mitochondrial electron transport chain (ETC). Electrons are transferred through Complex I to IV resulting in the leakage of electrons and finally the addition of electrons to oxygen (O_2). The proton (H^+) gradient generated by the electron transfer provides the energy needed to generate ATP (adenosine triphosphate) by the final Complex V or ATP synthase. CoQ = oxidized ubiquinone; $CoQH_2$ = reduced ubiquinol; e^- = electrons

$p67^{phox}$, as well as a small G protein, Rac, translocate to the cell membrane and associate with the two subunits of cytochrome b558 to generate O_2^- (Bedard and Krause 2007). The discovery of NOX in non-phagocytic cells led to the identification of six additional NOX isoforms (NOX1, 3–5, DUOX1, DUOX2) as well as two subunit isoforms, NOXO1 ($p47^{phox}$ isoform) and NOXA1 ($p67^{phox}$ isoform). The expression patterns of these isoforms are distinct and seem to be tissue-specific (Jiang et al. 2011). Interestingly, the NOX enzyme systems are reportedly only found in multicellular organisms (eukaryotes) (Panday et al. 2015; Lalucque and Silar 2003). While most of the NOX enzymes generate O_2^-, only NOX4, DUOX1, and DUOX2 generate H_2O_2 (MacFie et al. 2014; Yoshihara et al. 2012). The physiological function of NOX is the modulation of multiple redox-sensitive intracellular signaling pathways (Jiang et al. 2011).

4 ROS in Adipocyte Differentiation

4.1 Introduction

Elevated levels of ROS are associated with cancer, diabetes, cardiovascular, and other diseases (Alfadda and Sallam 2012a). Inflammation in chronic inflammatory diseases such as rheumatoid arthritis is also associated with ROS overproduction. The mechanisms of ROS production in these conditions are still uncertain. Obesity is also associated with chronic inflammation of adipose tissue and is strongly linked to other noncommunicable diseases including cancer, type 2 diabetes, and cardiovascular disease (Alfadda and Sallam 2012b; Roy et al. 2017). The worldwide prevalence of obesity and its comorbidities has reached alarming levels. This

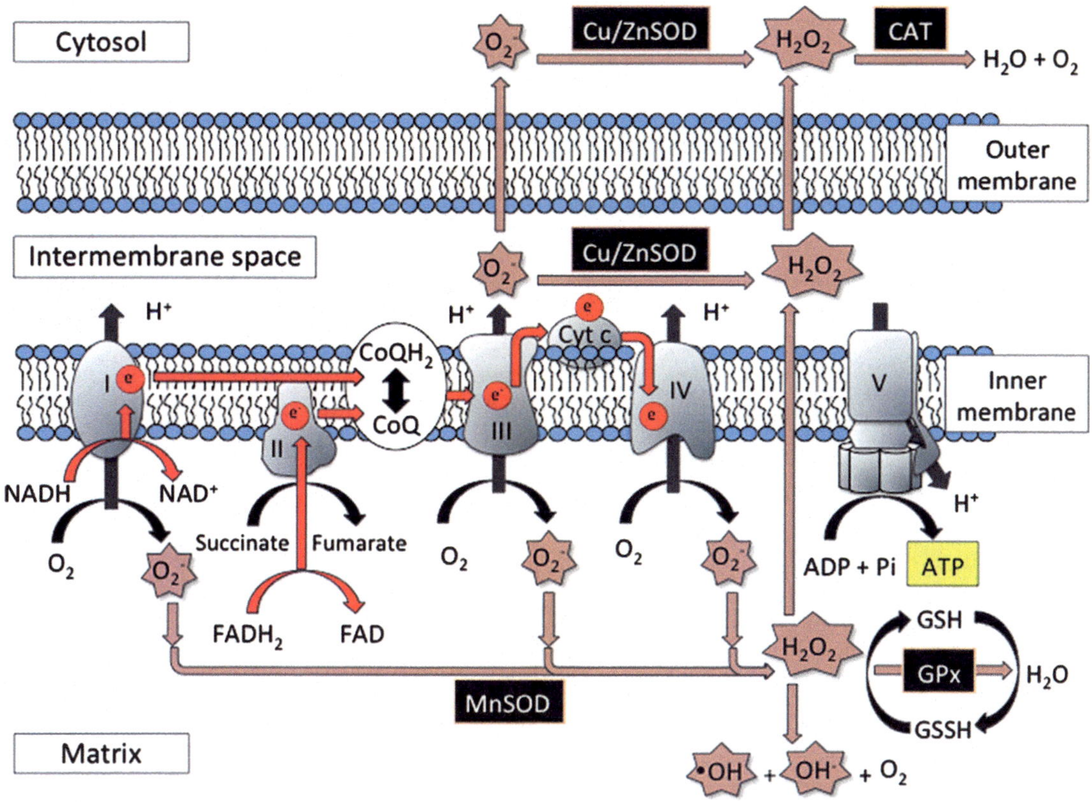

Fig. 3 An overview of the production of reactive oxygen species (ROS). Molecular O_2 may be reduced to superoxide anion radical (O_2^-), which can be further reduced to hydrogen peroxide (H_2O_2) either spontaneously or through the action of superoxide dismutase (SOD) enzymes (MnSOD and Cu/ZnSOD). The transition metals such as Fe^{2+} and Cu^+ catalyze the conversion of H_2O_2 to the hydroxyl radical (•OH) via the Fenton and Harber–Weiss reactions. The enzymes catalase (CAT) or glutathione peroxidase (GPx) detoxifies H_2O_2 by converting it to water (H_2O). GSSG = oxidized glutathione; GSH = reduced glutathione

Fig. 4 NADPH oxidase (NOX). The typical NADPH oxidase members are transmembrane proteins that transport electrons across biological membranes to reduce oxygen (O_2) to superoxide (O_2^-)

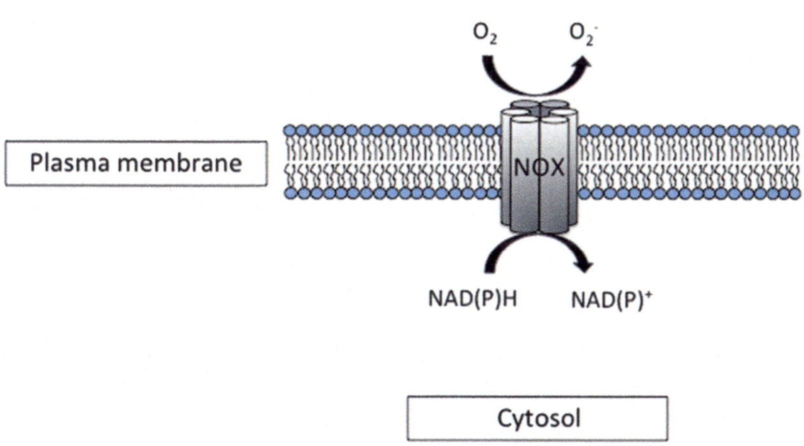

has led to an increase in studies designed to understand the biology of adipose tissue, the process of adipogenesis, as well as the investigation into molecules such as ROS, which could modulate the process. A better understanding of the mechanisms involved in obesity may not only provide insight into the development of novel treatment strategies to combat obesity but also the pathogenesis of other noncommunicable diseases.

4.2 Adipose-Derived Stromal Cells and their Ability to Differentiate into Adipocytes

Adipose tissue consists of various cell types, including adipocytes, preadipocytes, adipose-derived stem/stromal cells (ASCs), endothelial precursors, T regulatory cells, macrophages, smooth muscle cells, and pericytes (Riordan et al. 2009; Nguyen et al. 2016). Adipogenesis is the process by which MSCs differentiate into adipocytes, which forms the major cellular component of adipose tissue (Ambele et al. 2016; Ambele and Pepper 2017). Mesenchymal stem/stromal cells harvested from adipose tissue are referred to as adipose-derived stem/stromal cells (ASCs) (Bassi et al. 2012). The ability of ASCs to differentiate into adipocytes in vitro is often used as a model to study adipogenesis.

To understand the role of adipose tissue in obesity, it will be necessary to understand the process of adipogenesis in its entirety. However, the full process of adipogenesis is not well defined. In particular, there is a deficit in our understanding at a molecular and cellular level of the intermediate adipocyte subpopulations that exist from the stem cell to the mature adipocyte (Durandt et al. 2016; Rosen and MacDougald 2006). In an attempt to address the lack in knowledge regarding adipocyte subpopulations during adipocyte differentiation, Durandt et al. (2016) showed in a recent study that the cell surface expression of the fatty acid translocase (CD36/FAT) precedes lipid accumulation allowing the identification of at least three adipocyte subpopulations during adipocyte differentiation (Durandt et al. 2016). Adipogenesis is most often described to occur in two steps, namely, a determination phase and a terminal differentiation phase. The former consists of MSC commitment to differentiate into preadipocytes. These precursor cells do not possess defining morphological features or specific gene expression patterns that are distinct from their predecessors (Rosen and MacDougald 2006; Otto and Lane 2005), but have lost their ability to differentiate into other cell types and are committed to the adipocyte lineage (Otto and Lane 2005; Ali et al. 2013).

Several studies suggest that it is a prerequisite for preadipocytes to undergo growth arrest at the commitment phase for adipogenesis to proceed. Growth arrest, not necessarily cell-cell contact or confluence, is essential for differentiation to occur (Gregoire et al. 1998). Following growth arrest, upon reaching confluence and upon hormonal induction, it is reported that some cell lines, e.g., 3T3-L1, Ob17, and 3T3 F442A, undergo a mitotic clonal expansion phase prior to adipocyte differentiation. During the mitotic clonal expansion phase, the cells undergo one or two rounds of cell division prior to differentiation. On the other hand, studies using mouse CH3H10T1/2 cells and human preadipocytes have shown that the cells differentiate without post confluence mitosis (Gregoire et al. 1998).

The second phase in adipocyte differentiation is terminal differentiation into functional mature adipocytes with the accumulation of intracellular lipid droplets (Gregoire et al. 1998). Initially, lipid droplets appear in the cytoplasm near the periphery of preadipocytes, followed by the aggregation and enlargement of lipid droplets through fusion. The enlarged lipid droplets migrate more centrally in preadipocytes. A mature adipocyte is a terminally differentiated cell and contains a large unilocular lipid droplet. Lipid droplets are constitutively formed in adipocytes. However, non-adipocytes may also contain small mobile lipid droplets (Murphy et al. 2009).

Adipogenesis can be thought of as a transition through the up- and downregulation of thousands of different genes (Satish et al. 2015; Ambele et al. 2016). This makes it difficult to accurately discuss the progression of events in chronological order. Various positive and negative regulators are involved, and adipogenesis is a consequence of an equilibrium between these factors (Ambele and Pepper 2017). Positive regulators of adipogenesis include proliferator-activated receptor-gamma (PPARγ); CCAAT enhancer-binding protein (C/EBP) transcription factors; the Kruppel-like factor (KLF) family of C2H2 zinc finger proteins, e.g., KLF15, KLF5, KLF6, KLF9, and KROX20; sterol regulatory element-binding transcription factor 1 (SREBP1c); cyclic AMP response element-binding protein (CREB); the zinc finger protein 423 (ZFP423); signal transducer and activator of transcription-5a (STAT5a); endothelial PAS domain protein 1 (EPAS1); and brain and muscle ARNT-like protein 1 (BMAL1), among others (Rosen and MacDougald 2006; Moseti et al. 2016; Stephens 2012). Peroxisome proliferator-activated receptor-gamma (PPARγ) and C/EBP transcription factors play a key role in regulating terminal differentiation (Ambele et al. 2016; Ambele and Pepper 2017). There are three C/EBP transcription factors (C/EBPα, C/EBPβ, and C/EBPδ) that can form heterodimers with each other and that bind to C/EBP regulatory elements within promoters (Otto and Lane 2005). The first transcription factors to be induced in vitro upon exposure to glucocorticoids or insulin are C/EBPβ and C/EBPδ, which together activate C/EBPα expression. CCAAT enhancer-binding protein alpha is thought to play an important role in establishing terminally differentiated adipocytes (Samuelsson et al. 1991). Both C/EBPβ and C/EBPδ induce the upregulation of PPARγ, which together with C/EBPα coordinate the expression of various adipogenic-specific genes leading to the mature adipocyte phenotype (Fig. 5).

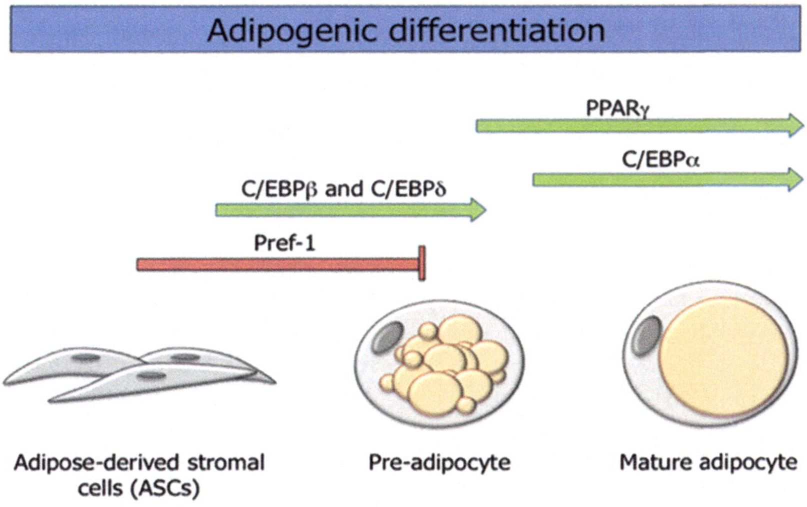

Fig. 5 Adipogenic differentiation. Adipose-derived stem/stromal cells (ASCs) can be committed to the adipogenic lineage and thus converted to preadipocytes, which further terminally differentiate into mature adipocytes. During the conversion of preadipocytes to adipocytes, growth arrest occurs. The subsequent activation of adipocyte genes by the transcription factors CCAAT/enhancer-binding protein (C/EBPα) and peroxisome proliferator-activated receptor-gamma (PPARγ) drives adipogenesis. Changes in morphology include the acquisition of a more spherical shape and the accumulation of lipid droplets

Among the repressors of adipogenesis is preadipocyte factor-1 (Pref-1). Downregulation of Pref-1 occurs during adipocyte differentiation (Wang et al. 2010a, b). Other negative regulators include KLF2, KLF7, GATA binding proteins 2 and 3 (GATA2 and GATA3), forkhead box O1 (FOXO1), forkhead box A2 (FOXA2), transcriptional coactivator with PDZ-binding motif (TAZ), and histone deacetylase sirtuin 1 (SIRT1). The hedgehog, bone morphogenic protein (BMP), and the β-catenin-dependent Wnt signaling pathways have also been implicated as negative regulators of adipogenesis (Stephens 2012; Rosen and MacDougald 2006; Moseti et al. 2016).

4.3 The Role of ROS in Adipocyte Differentiation

Several studies have demonstrated the important role of exogenous and intracellular generated ROS during adipogenesis using different experimental models including preadipocyte cell lines, primary bone marrow-derived MSCs (BM-MSCs), and ASCs isolated from the mouse, rat, and humans. The consensus from the majority of these studies is that ROS promote adipogenesis (Atashi et al. 2015; Wang et al. 2015; Lee et al. 2009). However, it is still not fully understood how molecules such as ROS affect this process. A better understanding of ROS during adipogenic differentiation could potentially lead to the development of new therapies for obesity and related comorbidities.

The in vitro adipogenic differentiation of 3T3-L1 cells has been shown to increase in the presence of exogenous H_2O_2 (Lee et al. 2009). The H_2O_2 associated increase in adipogenic differentiation was accompanied by increases in C/EBPβ and PPARγ expression. CCAAT enhancer-binding protein beta is thought to initiate mitotic clonal expansion of preadipocytes as well as cell cycle progression from the S to G2/M phase. Addition of antioxidants causes S phase arrest, thus preventing mitotic clonal expansion during differentiation. It was therefore suggested that ROS facilitate adipogenic differentiation by accelerating mitotic clonal expansion in preadipocytes (Lee et al. 2009). In vitro, a standard adipogenic induction cocktail, consisting of insulin, dexamethasone, indomethacin, and 3-isobutyl-1-methylxanthine (cocktail abbreviated as IDII), induced adipocyte differentiation in human ASCs, and this was accompanied by an increase in ROS production. IDII-induced adipogenic differentiation was inhibited by ROS scavengers such as N-acetyl-L-cysteine (NAC), SOD and CAT (Higuchi et al. 2013). Interestingly, several antioxidant enzymes, including superoxide dismutase 2 (SOD2), CAT, and GPx, were also found to be upregulated during induction of adipocyte differentiation using IDII. This IDII-induced upregulation of antioxidant expression was inhibited in FOXO1 transcription factor knockout experiments. FOXO transcription factors play an important role in maintaining cellular redox homeostasis and have been found to be significantly downregulated during IDII-induced adipogenesis (Higuchi et al. 2013). These findings suggest that a delicate balance between ROS production and endogenous antioxidant generation is maintained during adipogenesis. Several other studies, using mouse 3T3 preadipocyte cultures, also suggest that FOXO transcriptional factors are involved in the regulation of adipogenesis, but the findings are inconsistent and need further investigation. Nakae et al. (2003) showed that the overexpression of an activated form of FOXO1 inhibits adipogenesis by preventing clonal expansion of progenitor cells (Nakae et al. 2003). Contrary to these findings, Munekata and Sakamoto (2009) reported that FOXO1 silencing results in a decrease in adipogenic differentiation (Munekata and Sakamoto 2009).

Obesity is associated with increased infiltration of classical pro-inflammatory (M1) macrophages in the adipose tissue (Weisberg et al. 2003; Thomas and Apovian 2017). One of the key functions of M1 macrophages is the production of ROS and nitric oxide (NO) by NOX and inducible nitric oxide synthase (iNOS), respectively. The overproduction of ROS and NO in turn mediates the release

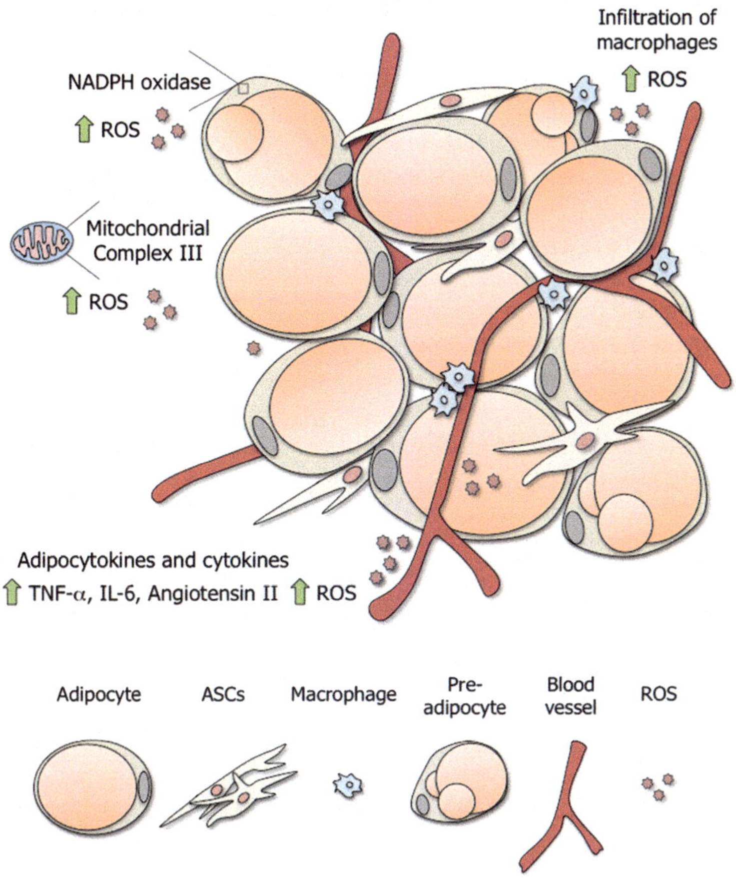

Fig. 6 Schematic illustrating how ROS production is increased in accumulating adipose tissue and how this ROS may contribute to adipogenesis. Reactive oxygen species are derived from multiple sources such as NADPH oxidase (NOX) and mitochondrial Complex III

of pro-inflammatory cytokines. This highlights another potential link between ROS and adipocyte differentiation and suggests that several active cellular mediators such as chemokines, cytokines, and adipokines contribute to the chronic inflammatory state of adipose tissue. This in turn contributes to the overproduction of ROS, causing systemic oxidative stress (Alfadda and Sallam 2012b). Figure 6 summarizes how ROS production is increased in adipose tissue.

Various studies suggest that NOX is one of the main sources of ROS production in adipose tissue. Two independent studies using high-fat diet fed murine experimental models have shown that obesity is associated with the activation of NOX (Chen and Stinnett 2008; Thomas and Apovian 2017). Treating obese mice with the NOX inhibitor apocynin resulted in a significant decrease in ROS production in adipose tissue as well as a decrease in the expression of the pro-inflammatory cytokine tumor necrosis factor alpha (TNFα) (Furukawa et al. 2004). Several in vitro studies also suggest that preadipocytes and adipocytes produce ROS through insulin-induced NOX-associated pathways (Liu et al. 2012a; Wang et al. 2015; Krieger-Brauer et al. 1997). ROS generated during the adipogenic differentiation of 3T3-L1 preadipocytes and ASCs were suppressed by NOX inhibitors (Liu et al. 2012a; Kanda et al. 2011). Other studies have shown that gene silencing of NOX4 inhibited insulin-induced terminal differentiation of 3T3-L1 cells, suggesting a positive role for NOX4 in promoting adipogenesis through insulin signaling (Schroder et al. 2009; Mahadev et al. 2004; Kanda et al. 2011; Mitchell et al.

2003; Mouche et al. 2007). NADPH oxidase 4 impairs insulin-induced proliferation but facilitates insulin-induced differentiation of preadipocytes by controlling the expression of MAP kinase phosphatase-1 (MKP-1), which limits ERK1/2 signaling, thereby acting as a switch from proliferation to differentiation in response to insulin (Schroder et al. 2009). Furthermore, overexpression of NOX4 in ASCs and exogenous application of H_2O_2, enhanced adipogenic differentiation (Higuchi et al. 2013), thereby supporting the idea of a positive influence of ROS in adipogenesis. Mouche et al. (2007) showed that NOX4 is highly expressed in primary preadipocytes isolated from murine brown and white adipose tissue (BAT and WAT). As confirmation, these investigators also showed that NOX4 is located in vivo in the stromal vascular fraction of adipose tissue rather than in the mature adipocyte fraction. In addition, systemic administration of a low molecular weight O_2^- scavenging agent, Tempol, in C3H mice through either drinking H_2O or food, prevented mice from becoming obese and also reduced their leptin levels (Mitchell et al. 2003). This complements in vitro findings on a positive role for ROS in adipogenesis. Contrary to the above-mentioned findings, Li and colleagues (2012) reported that mice deficient in NOX4 showed accelerated development of obesity and insulin resistance (Li et al. 2012), which brings into question the exact role of NOX4 in adipogenesis in vivo. Kanda et al. (2011) suggested in their study that ROS production by NOX4 promotes adipocyte differentiation through CREB (Kanda et al. 2011). These investigators showed that H_2O_2 induced the activation of CREB and that treatment of the 10T1/2 multipotent mouse embryonic cell line with the pharmacological antioxidant NAC blocked the expression of CREB and consequently inhibited adipogenic differentiation [6]. Cyclic AMP response element-binding protein is an early regulator of adipogenesis and is reported to regulate C/EBPβ expression, which when upregulated leads to the upregulation of C/EBPα and PPARγ, two key regulatory transcriptional factors in the adipogenic process. Recently, both isoforms of NOS, endothelial nitric oxide (eNOS) and iNOS, were detected in differentiating preadipocytes derived from rat WAT, with an approximately 50% increase in NOS over basal level observed in preadipocytes on the first two days of adipogenic differentiation (Yan et al. 2002). In this study, eNOS was shown to be the major isoform promoting adipogenesis as iNOS was found to have little effect on total NO production and preadipocyte differentiation. Currently, the role of other NOX isoforms (NOX1, NOX2, NOX3, and NOX5) in adipogenesis is unknown and needs to be investigated.

Although most studies suggest that NOX is mainly responsible for ROS overproduction during adipogenesis, a few studies have also investigated the role of mitochondrial ROS in this context. In one of the first studies that investigated the effect of mitochondrial ROS on adipogenesis, Carriere and colleagues showed that the inhibition of mitochondrial ETC Complex I and V leads to an increase in mitochondrial ROS production which prevents 3T3-L1 proliferation (Carriere et al. 2003). A few years later, Tormos and colleagues showed, using primary BM-MSCs, that an increase in mitochondrial metabolism is essential for and promotes adipocyte differentiation through the overproduction of ROS via mitochondrial Complex III (Tormos et al. 2011). Mitochondrial Complex III-derived ROS promotes adipogenic differentiation via the upregulation of C/EBPα and PPARγ (Tormos et al. 2011). This finding was confirmed using ASCs in which partial suppression of adipogenic differentiation was observed after treatment with rotenone (Liu et al. 2012a). Rotenone prevents mitochondrial ROS production by inhibiting the electron transport function of Complex I (Lindahl and Oberg 1960; Li et al. 2003). These findings suggest a complex role of mitochondria-derived ROS in adipogenesis.

It is suggested that ROS promote adipogenesis via two closely related mechanisms: (a) the promotion of adipocyte differentiation and (b) intracellular lipid accumulation. The enhancement of adipogenesis with increased NOS production was evidenced by increased lipoprotein lipase (LPL) expression and accelerated triacylglycerol (TG) accumulation in differentiating adipocytes (Yan et al. 2002). The accumulation of intracellular ROS (O_2^-) and intracellular lipid droplets in OP9 mouse stromal preadipocytes was shown to increase with adipogenic differentiation when compared to undifferentiated cells. The presence of an antioxidant led to a decrease in differentiation-dependent lipid droplets and intracellular ROS accumulation in OP9 cells during adipocyte differentiation (Saitoh et al. 2010). Not surprisingly, these studies reveal a positive correlation between increased endogenous ROS production and lipid droplet accumulation during adipocyte differentiation.

Irrespective of the intracellular source of ROS, studies using different experimentally designed approaches including manipulation of gene expression, treatment with antioxidant agents, as well as the direct measurement of intracellular ROS levels have all confirmed a close link between adipocyte differentiation and ROS (Younce and Kolattukudy 2012; Liu et al. 2012b; Nam et al. 2012; Hou et al. 2012; Imhoff and Hansen 2011; Saitoh et al. 2010; Samuni et al. 2010; Vigilanza et al. 2011; Atashi et al. 2015).

4.4 ROS Scavengers, Adipocytes and Adipogenesis

Under normal physiological conditions, ROS production in cells is countered by antioxidant defense mechanisms (ROS scavengers), which can either be enzymatic or nonenzymatic. Reactive oxygen species production during the process of adipogenesis is also accompanied by induction of endogenous antioxidant defense mechanisms. These antioxidant mechanisms scavenge excess ROS to prevent oxidative stress and to maintain ROS levels necessary for cellular homeostasis.

An increase in ROS production during the first seven days of human ASC adipogenic differentiation was accompanied by an increase in the activities of SOD and CAT antioxidant enzymatic systems and a reduction in nonprotein thiols such as glutathione, thereby indicating the presence of a ROS scavenging effect during adipogenesis (Drehmer et al. 2016). Adipose tissue samples from obese rats and mice showed activity of SOD, CAT, and GPx antioxidant enzymatic systems. The decrease in the redox state of glutathione by antioxidant treatment was shown to promote TG accumulation in preadipocytes in vitro (Galinier et al. 2006). The markedly increased ROS production during 3T3-L1 cell differentiation into adipocytes was suppressed by diphenyleneiodonium (DPI) and apocynin (a NOX-specific inhibitor), which are two structurally unrelated NOX inhibitors, and by the general antioxidant NAC (Furukawa et al. 2004). Six weeks of apocynin treatment significantly reduced lipid peroxidation and H_2O_2 production in WAT in a KKAy mouse model, thereby counteracting the positive effect of NOX-specific ROS production on adipogenesis (Furukawa et al. 2004). In addition, the downregulation of glutathione S-transferase A4 (GSTA4) (enzyme responsible for lipid aldehyde detoxification) in cultured 3T3-L1 adipocytes resulted in increased mitochondrial ROS production, lipolysis, and protein carbonylation (Curtis et al. 2010) indicating the significance of GSTA4 as a ROS scavenger in adipocytes.

Treatment of 3T3-L1 cells with the flavonoid, quercetin, with reported antioxidant properties, using a serum concentration range of 0.1–10 μM, inhibited TG accumulation in developing adipocytes and reduced the TG content in mature adipocytes. Furthermore, this treatment downregulated the expression of C/EBPβ, PPARγ, SREBP1c, and LPL genes, thus inhibiting adipogenesis (Eseberri et al. 2015). This is in accordance with another study that reported that quercetin as well as resveratrol at 25 μM individually inhibited lipid accumulation in 3T3-L1 preadipocytes (Yang et al. 2008). Interestingly, when quercetin and resveratrol were combined at 25 μM, the expression of key adipogenic transcription factors such as PPARγ1, PPARγ2, and C/EBPα was significantly suppressed (Yang et al. 2008). This suggests that ROS scavengers can work synergistically to inhibit adipogenesis. However, individually, quercetin and resveratrol had no effect on PPARγ2 and C/EBPα but reduced the expression of PPARγ1 (Yang et al. 2008). Genistein is an isoflavone that has been shown to inhibit the accumulation of lipid as well as to decrease the nonesterified fatty acid (NEFA) content of 3T3-L1 preadipocytes on day 6 of adipogenic differentiation (Zhang et al. 2009). The suppression of 3T3-L1 differentiation by genistein is suggested to be through multiple signaling pathways, which includes the janus-activated kinase and p38 pathways (Zhang et al. 2009). Catechins, which are antioxidant flavonoids from tea, have been shown to inhibit adipogenesis in 3T3-L1 adipocytes by inhibiting intracellular lipid accumulation and glycerol-3-phosphate dehydrogenase activity in the differentiating cells. This is accompanied by the downregulation of PPARγ2, C/EBPα, and GLUT4 (Furuyashiki et al. 2004). The antioxidant glutathione precursor, NAC, at 5 mM significantly decreased the accumulation of triglyceride in differentiating mouse embryonic fibroblasts. N-acetyl-L-cysteine was also reported to significantly inhibit the mitotic clonal expansion of these cells (Pieralisi et al. 2016), thereby negatively affecting adipogenesis. Bixin, β-carotene, and lycopene have been shown to inhibit intracellular lipid accumulation in 3T3-L1 cells during adipogenic differentiation, which is accompanied by the downregulation of PPARγ, fatty acid-binding protein 4 (FABP4), and leptin protein expression. β-carotene had the strongest inhibitory effect (Zhao et al. 2017).

5 Clinical Implications

The incidence of obesity and its related metabolic disorders is on a steady increase worldwide. The development of any preventive or therapeutic strategy requires a greater understanding of its etiology. The existence of a state of oxidative stress in obesity has been linked to some of its related pathophysiological conditions. This has contributed to the understanding of some associated metabolic disorders and their attenuation by pharmaceuticals in rodent studies (Le Lay et al. 2014). Oxidative stress in adipose tissue results from an imbalance between ROS production and neutralization by antioxidants and has been linked to adipocyte dysfunction. Reactive oxygen species such as NOX4-generated H_2O_2 are important modulators of adipogenesis and adipose tissue function, which is key in health and disease (Tormos et al. 2011; Kanda et al. 2011; Higuchi et al. 2013) since impairment of adipocyte differentiation and metabolism is said to precede WAT dysfunction. Studies on C57BL/6 mice in which NOX4 had been specifically deleted in adipocytes reveal a role for NOX4-derived ROS in the onset of insulin resistance and adipose tissue inflammation (Den Hartigh et al. 2016). Between 10 and 20% of type 2 diabetes are considered to have a familial component (Drong et al. 2012), while obesity remains the common and main risk factor for the development type 2 diabetes (Ambele et al. 2016). The growth of adipocytes beyond a critical volume (the tolerated adipocyte volume which does not compromise its functions, depends on individual parameters that are not yet fully understood) is linked to functional impairment and an increased risk of developing type 2 diabetes (Guilherme et al. 2008; Cotillard et al. 2014). The link between adipocyte dysfunction and type 2 diabetes has been called adipose tissue expandability, which states that it is the failure of adipose tissue expansion capacity, rather than obesity per se, which is the key factor linking positive energy balance and type 2 diabetes (Virtue and Vidal-Puig 2010). This suggests an important role for adipocyte formation/adipogenesis, which is under the influence of ROS, in the development of type 2 diabetes. In addition, increased lipolysis, impairment of insulin-stimulated glucose uptake, leptin secretion, and increased secretion of pro-inflammatory cytokines such as interleukin 6 (IL-6), IL-8, and TNFα, and reduced secretion of adiponectin and IL-10, have been associated with adipocyte hypertrophy (Castro et al. 2016). The secretome of dysfunctional WAT has been shown to contribute locally and systemically to a chronic pro-inflammatory state, which increases insulin resistance through an imbalanced secretion of insulin sensitizing (adiponectin) and inhibiting (TNFα, C-C motif chemokine 2 precursor; CCL2) adipokines (Castro et al. 2016). Macrophage infiltration has been observed in low-grade inflammation of WAT in obese subjects, which correlates with adipocyte mean size and body mass (Bremer and Jialal 2013; Weisberg et al. 2003). Han (2016) has suggested that NOX4-derived ROS from adipocytes in the early stages of obesity provoke the onset of insulin resistance and initiate the recruitment of immune cells in adipose tissue (Han 2016). NOX4-derived ROS from infiltrating immune cells such as macrophages worsen insulin resistance and adipose tissue inflammation at the intermediate stages of obesity, while mitochondria-derived ROS maintain insulin resistance and adipose tissue inflammation at the late stages of obesity (Han 2016).

Overall, ROS is important for the differentiation of preadipocytes into adipocytes; however, an increase in ROS in mature adipocytes leads to adipocyte dysfunction, which causes oxidative stress and inflammation within and around WAT, thereby contributing to the development of type 2 diabetes and other metabolic disorders. Therefore, the pathophysiological role of ROS in biological processes, including adipogenesis, adipose tissue function, obesity, obesity-related inflammation, and obesity-related comorbidities such as type 2 diabetes, highlights the importance of ROS as clinically relevant molecules. Reactive oxygen species could either serve as biomarkers or targets in combating these related diseases.

6 Conclusion

Recent advances in research on ROS and adipogenesis has elicited much interest in the fields of obesity and other metabolic diseases. Understanding how ROS influence adipogenic differentiation and adipose tissue function in vivo nonetheless remains a challenge. Both extra- and intracellular ROS promote adipocyte differentiation at multiple stages. However, to date, research findings are unable to fully explain the exact source(s) of ROS that initiate and/or drive adipogenic differentiation in vivo. Clarity on the mechanistic interplay between the source/type of ROS and ROS inhibitors regulating adipocyte differentiation in vivo is likely to open new avenues for combating obesity and other related metabolic disease processes.

Acknowledgements This research was funded by the South African Medical Research Council in terms of the SAMRC's Flagship Award Project SAMRC-RFA-UFSP-01-2013/STEM CELLS, the SAMRC Extramural Unit for Stem Cell Research and Therapy and the Institute for Cellular and Molecular Medicine of the University of Pretoria.

Conflicts of Interest Statement The authors have no conflicts of interest to declare.

Author Contributions Danielle de Villiers drafted the first version of the manuscript with input from Marnie Potgieter and Michael Pepper. Melvin Ambele, Chrisna Durandt, and Ladislaus Adam provided intellectual input and contributed to the writing of the manuscript. All authors vetted and approved the final version of the manuscript. Michael Pepper conceived the project.

References

Alfadda, A. A., & Sallam, R. M. (2012a). Reactive oxygen species in health and disease. *Journal of Biomedicine and Biotechnology, 2012*, 14. https://doi.org/10.1155/2012/936486.

Alfadda, A. A., & Sallam, R. M. (2012b). Reactive oxygen species in health and disease. *Journal of Biomedicine & Biotechnology, 2012*, 936486. https://doi.org/10.1155/2012/936486.

Ali, A. T., Hochfeld, W. E., Myburgh, R., & Pepper, M. S. (2013). Adipocyte and adipogenesis. *European Journal of Cell Biology, 92*(6–7), 229–236. https://doi.org/10.1016/j.ejcb.2013.06.001.

Ambele, M. A., & Pepper, M. S. (2017). Identification of transcription factors potentially involved in human adipogenesis in vitro. *Molecular Genetics & Genomic Medicine:n/a-n/a., 5*, 210. https://doi.org/10.1002/mgg3.269.

Ambele, M. A., Dessels, C., Durandt, C., & Pepper, M. S. (2016). Genome-wide analysis of gene expression during adipogenesis in human adipose-derived stromal cells reveals novel patterns of gene expression during adipocyte differentiation. *Stem Cell Research, 16*(3), 725–734. https://doi.org/10.1016/j.scr.2016.04.011.

Atashi, F., Modarressi, A., & Pepper, M. S. (2015). The role of reactive oxygen species in mesenchymal stem cell Adipogenic and osteogenic differentiation: A review. *Stem Cells and Development, 24*(10), 1150–1163. https://doi.org/10.1089/scd.2014.0484.

Babior, B. M., Kipnes, R. S., & Curnutte, J. T. (1973). Biological defense mechanisms. The production by leukocytes of superoxide, a potential bactericidal agent. *The Journal of Clinical Investigation, 52*(3), 741–744. https://doi.org/10.1172/jci107236.

Bassi, G., Pacelli, L., Carusone, R., Zanoncello, J., & Krampera, M. (2012). Adipose-derived stromal cells (ASCs). *Transfusion and Apheresis Science : Official Journal of the World Apheresis Association : Official Journal of the European Society for Haemapheresis, 47*(2), 193–198. https://doi.org/10.1016/j.transci.2012.06.004.

Bedard, K., & Krause, K. H. (2007). The NOX family of ROS-generating NADPH oxidases: Physiology and pathophysiology. *Physiological Reviews, 87*(1), 245–313. https://doi.org/10.1152/physrev.00044.2005.

Bekker, A., Holland, H. D., Wang, P. L., Rumble, D., 3rd, Stein, H. J., Hannah, J. L., Coetzee, L. L., & Beukes, N. J. (2004). Dating the rise of atmospheric oxygen. *Nature, 427*(6970), 117–120. https://doi.org/10.1038/nature02260.

Bremer, A. A., & Jialal, I. (2013). Adipose tissue dysfunction in nascent metabolic syndrome. *Journal of Obesity, 2013*, 393192. https://doi.org/10.1155/2013/393192.

Carriere, A., Fernandez, Y., Rigoulet, M., Penicaud, L., & Casteilla, L. (2003). Inhibition of preadipocyte proliferation by mitochondrial reactive oxygen species. *FEBS Letters, 550*(1–3), 163–167.

Castro, J. P., Grune, T., & Speckmann, B. (2016). The two faces of reactive oxygen species (ROS) in adipocyte function and dysfunction. *Biological Chemistry, 397* (8), 709–724. https://doi.org/10.1515/hsz-2015-0305.

Chen, Y., & Junger, W. G. (2012). Measurement of oxidative burst in neutrophils. *Methods in Molecular Biology (Clifton, NJ), 844*, 115–124. https://doi.org/10.1007/978-1-61779-527-5_8.

Chen, J.-X., & Stinnett, A. (2008). Ang-1 gene therapy inhibits hypoxia-inducible factor-1α (HIF-1α)-Prolyl-4-Hydroxylase-2, stabilizes HIF-1α expression, and normalizes immature vasculature in db/db mice. *Diabetes, 57*(12), 3335–3343. https://doi.org/10.2337/db08-0503.

Commoner, B., Townsend, J., & Pake, G. E. (1954). Free radicals in biological materials. *Nature, 174*(4432), 689–691.

Cotillard, A., Poitou, C., Torcivia, A., Bouillot, J. L., Dietrich, A., Kloting, N., Gregoire, C., Lolmede, K., Bluher, M., & Clement, K. (2014). Adipocyte size threshold matters: Link with risk of type 2 diabetes and improved insulin resistance after gastric bypass. *The Journal of Clinical Endocrinology and Metabolism, 99*(8), E1466–E1470. https://doi.org/10.1210/jc.2014-1074.

Curtis, J. M., Grimsrud, P. A., Wright, W. S., Xu, X., Foncea, R. E., Graham, D. W., Brestoff, J. R., Wiczer, B. M., Ilkayeva, O., Cianflone, K., Muoio, D. E., Arriaga, E. A., & Bernlohr, D. A. (2010). Downregulation of adipose glutathione S-transferase A4 leads to increased protein carbonylation, oxidative stress, and mitochondrial dysfunction. *Diabetes, 59* (5), 1132–1142. https://doi.org/10.2337/db09-1105.

D'Autreaux, B., & Toledano, M. B. (2007). ROS as signalling molecules: Mechanisms that generate specificity in ROS homeostasis. *Nature Reviews Molecular Cell Biology, 8*(10), 813–824. https://doi.org/10.1038/nrm2256.

Den Hartigh, L. J., Omer, M., Goodspeed, L., Wang, S., Wietecha, T., O'Brien, K. D., & Han, C. Y. (2016). Adipocyte-specific deficiency of NADPH oxidase 4 delays the onset of insulin resistance and attenuates adipose tissue inflammation in obesity. *Arteriosclerosis, Thrombosis, and Vascular Biology, 37*(3), 466–475.

Drehmer, D. L., de Aguiar, A. M., Brandt, A. P., Petiz, L., Cadena, S. M., Rebelatto, C. K., Brofman, P. R., Filipak Neto, F., Dallagiovanna, B., & Abud, A. P. (2016). Metabolic switches during the first steps of adipogenic stem cells differentiation. *Stem Cell Research, 17*(2), 413–421. https://doi.org/10.1016/j.scr.2016.09.001.

Droge, W. (2002). Free radicals in the physiological control of cell function. *Physiological Reviews, 82*(1), 47–95. https://doi.org/10.1152/physrev.00018.2001.

Drong, A. W., Lindgren, C. M., & McCarthy, M. I. (2012). The genetic and epigenetic basis of type 2 diabetes and obesity. *Clinical Pharmacology and Therapeutics, 92*(6), 707–715. https://doi.org/10.1038/clpt.2012.149.

Drose, S., & Brandt, U. (2008). The mechanism of mitochondrial superoxide production by the cytochrome bc1 complex. *The Journal of Biological Chemistry, 283*(31), 21649–21654. https://doi.org/10.1074/jbc.M803236200.

Durandt, C., van Vollenstee, F. A., Dessels, C., Kallmeyer, K., de Villiers, D., Murdoch, C., Potgieter, M., & Pepper, M. S. (2016). Novel flow cytometric approach for the detection of adipocyte sub-populations during adipogenesis. *Journal of Lipid Research, 57*(4), 729–742. https://doi.org/10.1194/jlr.D065664.

Eseberri, I., Miranda, J., Lasa, A., Churruca, I., & Portillo, M. P. (2015). Doses of quercetin in the range of serum concentrations exert delipidating effects in 3T3-L1 preadipocytes by acting on different stages of adipogenesis, but not in mature adipocytes. *Oxidative Medicine and Cellular Longevity, 2015*, 480943. https://doi.org/10.1155/2015/480943.

Finkel, T. (2011). Signal transduction by reactive oxygen species. *The Journal of Cell Biology, 194*(1), 7–15. https://doi.org/10.1083/jcb.201102095.

Furukawa, S., Fujita, T., Shimabukuro, M., Iwaki, M., Yamada, Y., Nakajima, Y., Nakayama, O., Makishima, M., Matsuda, M., & Shimomura,

I. (2004). Increased oxidative stress in obesity and its impact on metabolic syndrome. *The Journal of Clinical Investigation, 114*(12), 1752–1761. https://doi.org/10.1172/jci21625.

Furuyashiki, T., Nagayasu, H., Aoki, Y., Bessho, H., Hashimoto, T., Kanazawa, K., & Ashida, H. (2004). Tea catechin suppresses adipocyte differentiation accompanied by down-regulation of PPARgamma2 and C/EBPalpha in 3T3-L1 cells. *Bioscience, Biotechnology, and Biochemistry, 68*(11), 2353–2359. https://doi.org/10.1271/bbb.68.2353.

Galinier, A., Carriere, A., Fernandez, Y., Carpene, C., Andre, M., Caspar-Bauguil, S., Thouvenot, J. P., Periquet, B., Penicaud, L., & Casteilla, L. (2006). Adipose tissue proadipogenic redox changes in obesity. *The Journal of Biological Chemistry, 281*(18), 12682–12687. https://doi.org/10.1074/jbc.M506949200.

Gregoire, F. M., Smas, C. M., & Sul, H. S. (1998). Understanding adipocyte differentiation. *Physiological Reviews, 78*(3), 783–809.

Guilherme, A., Virbasius, J. V., Puri, V., & Czech, M. P. (2008). Adipocyte dysfunctions linking obesity to insulin resistance and type 2 diabetes. *Nature Reviews Molecular Cell Biology, 9*(5), 367–377. https://doi.org/10.1038/nrm2391.

Han, C. Y. (2016). Roles of reactive oxygen species on insulin resistance in adipose tissue. *Diabetes & Metabolism Journal, 40*(4), 272–279. https://doi.org/10.4093/dmj.2016.40.4.272.

Harman, D. (1956). Aging: A theory based on free radical and radiation chemistry. *Journal of Gerontology, 11*(3), 298–300.

Higuchi, M., Dusting, G. J., Peshavariya, H., Jiang, F., Hsiao, S. T., Chan, E. C., & Liu, G. S. (2013). Differentiation of human adipose-derived stem cells into fat involves reactive oxygen species and Forkhead box O1 mediated upregulation of antioxidant enzymes. *Stem Cells and Development, 22*(6), 878–888. https://doi.org/10.1089/scd.2012.0306.

Hinkle, P. C., Butow, R. A., Racker, E., & Chance, B. (1967). Partial resolution of the enzymes catalyzing oxidative phosphorylation. XV. Reverse electron transfer in the flavin-cytochrome beta region of the respiratory chain of beef heart submitochondrial particles. *The Journal of Biological Chemistry, 242*(22), 5169–5173.

Holzerova, E., & Prokisch, H. (2015). Mitochondria: Much ado about nothing? How dangerous is reactive oxygen species production? *The International Journal of Biochemistry & Cell Biology, 63*, 16–20. https://doi.org/10.1016/j.biocel.2015.01.021.

Hou, Y., Xue, P., Bai, Y., Liu, D., Woods, C. G., Yarborough, K., Fu, J., Zhang, Q., Sun, G., Collins, S., Chan, J. Y., Yamamoto, M., Andersen, M. E., & Pi, J. (2012). Nuclear factor erythroid-derived factor 2-related factor 2 regulates transcription of CCAAT/enhancer-binding protein beta during adipogenesis. *Free Radical Biology & Medicine, 52*(2), 462–472. https://doi.org/10.1016/j.freeradbiomed.2011.10.453.

Imhoff, B. R., & Hansen, J. M. (2011). Differential redox potential profiles during adipogenesis and osteogenesis. *Cellular & Molecular Biology Letters, 16*(1), 149–161. https://doi.org/10.2478/s11658-010-0042-0.

Iyer, G. Y. N., Islam, M. F., & Quastel, J. H. (1961). Biochemical aspects of phagocytosis. *Nature, 192*(4802), 535–541.

Jiang, F., Zhang, Y., & Dusting, G. J. (2011). NADPH oxidase-mediated redox signaling: Roles in cellular stress response, stress tolerance, and tissue repair. *Pharmacological Reviews, 63*(1), 218–242. https://doi.org/10.1124/pr.110.002980.

Jiang, Y., Jo, A. Y., & Graff, J. M. (2012). SnapShot: Adipocyte life cycle. *Cell, 150*(1), 234–234.e232. https://doi.org/10.1016/j.cell.2012.06.022.

Kanda, Y., Hinata, T., Kang, S. W., & Watanabe, Y. (2011). Reactive oxygen species mediate adipocyte differentiation in mesenchymal stem cells. *Life Sciences, 89*(7–8), 250–258. https://doi.org/10.1016/j.lfs.2011.06.007.

Kawagishi, H., & Finkel, T. (2014). Unraveling the truth about antioxidants: ROS and disease: Finding the right balance. *Nature Medicine, 20*(7), 711–713. https://doi.org/10.1038/nm.3625.

Krieger-Brauer, H. I., Medda, P. K., & Kather, H. (1997). Insulin-induced activation of NADPH-dependent H2O2 generation in human adipocyte plasma membranes is mediated by Galphai2. *The Journal of Biological Chemistry, 272*(15), 10135–10143.

Lalucque, H., & Silar, P. (2003). NADPH oxidase: An enzyme for multicellularity? *Trends in Microbiology, 11*(1), 9–12.

Le Lay, S., Simard, G., Martinez, M. C., & Andriantsitohaina, R. (2014). Oxidative stress and metabolic pathologies: From an adipocentric point of view. *Oxidative Medicine and Cellular Longevity, 2014*, 18. https://doi.org/10.1155/2014/908539.

Lee, H., Lee, Y. J., Choi, H., Ko, E. H., & J-w, K. (2009). Reactive oxygen species facilitate adipocyte differentiation by accelerating mitotic clonal expansion. *The Journal of Biological Chemistry, 284*(16), 10601–10609. https://doi.org/10.1074/jbc.M808742200.

Lefterova, M. I., & Lazar, M. A. (2009). New developments in adipogenesis. *Trends in Endocrinology and Metabolism: TEM, 20*(3), 107–114. https://doi.org/10.1016/j.tem.2008.11.005.

Li, N., Ragheb, K., Lawler, G., Sturgis, J., Rajwa, B., Melendez, J. A., & Robinson, J. P. (2003). Mitochondrial complex I inhibitor rotenone induces apoptosis through enhancing mitochondrial reactive oxygen species production. *The Journal of Biological Chemistry, 278*(10), 8516–8525. https://doi.org/10.1074/jbc.M210432200.

Li, Y., Mouche, S., Sajic, T., Veyrat-Durebex, C., Supale, R., Pierroz, D., Ferrari, S., Negro, F., Hasler, U., Feraille, E., Moll, S., Meda, P., Deffert, C., Montet, X., Krause, K. H., & Szanto, I. (2012). Deficiency in the NADPH oxidase 4 predisposes towards diet-induced obesity. *International Journal of Obesity, 36* (12), 1503–1513. https://doi.org/10.1038/ijo.2011.279.

Lindahl, P. E., & Oberg, K. E. (1960). Mechanism of the physiological action of rotenone. *Nature, 187*, 784.

Liu, G.-S., Chan, E. C., Higuchi, M., Dusting, G. J., & Jiang, F. (2012a). Redox mechanisms in regulation of adipocyte differentiation: Beyond a general stress response. *Cell, 1*(4), 976–993. https://doi.org/10.3390/cells1040976.

Liu, H., Yang, X., Zhang, Y., Dighe, A., Li, X., & Cui, Q. (2012b). Fullerol antagonizes dexamethasone-induced oxidative stress and adipogenesis while enhancing osteogenesis in a cloned bone marrow mesenchymal stem cell. *Journal of Orthopaedic Research: Official Publication of the Orthopaedic Research Society, 30*(7), 1051–1057. https://doi.org/10.1002/jor.22054.

Loschen, G., Azzi, A., Richter, C., & Flohe, L. (1974). Superoxide radicals as precursors of mitochondrial hydrogen peroxide. *FEBS Letters, 42*(1), 68–72.

MacFie, T. S., Poulsom, R., Parker, A., Warnes, G., Boitsova, T., Nijhuis, A., Suraweera, N., Poehlmann, A., Szary, J., Feakins, R., Jeffery, R., Harper, R. W., Jubb, A. M., Lindsay, J. O., & Silver, A. (2014). DUOX2 and DUOXA2 form the predominant enzyme system capable of producing the reactive oxygen species H2O2 in active ulcerative colitis and are modulated by 5-aminosalicylic acid. *Inflammatory Bowel Diseases, 20*(3), 514–524. https://doi.org/10.1097/01.MIB.0000442012.45038.0e.

Mahadev, K., Motoshima, H., Wu, X., Ruddy, J. M., Arnold, R. S., Cheng, G., Lambeth, J. D., & Goldstein, B. J. (2004). The NAD(P)H oxidase homolog Nox4 modulates insulin-stimulated generation of H2O2 and plays an integral role in insulin signal transduction. *Molecular and Cellular Biology, 24*(5), 1844–1854.

McCord, J. M., & Fridovich, I. (1969). Superoxide dismutase. An enzymic function for erythrocuprein (hemocuprein). *The Journal of Biological Chemistry, 244*(22), 6049–6055.

Meier, B., Cross, A. R., Hancock, J. T., Kaup, F. J., & Jones, O. T. (1991). Identification of a superoxide-generating NADPH oxidase system in human fibroblasts. *The Biochemical Journal, 275*(Pt 1), 241–245.

Mitchell, J. B., Xavier, S., DeLuca, A. M., Sowers, A. L., Cook, J. A., Krishna, M. C., Hahn, S. M., & Russo, A. (2003). A low molecular weight antioxidant decreases weight and lowers tumor incidence. *Free Radical Biology & Medicine, 34*(1), 93–102.

Moseti, D., Regassa, A., & Kim, W. K. (2016). Molecular regulation of Adipogenesis and potential anti-Adipogenic bioactive molecules. *International Journal of Molecular Sciences, 17*(1), 124. https://doi.org/10.3390/ijms17010124.

Mouche, S., Mkaddem, S. B., Wang, W., Katic, M., Tseng, Y. H., Carnesecchi, S., Steger, K., Foti, M., Meier, C. A., Muzzin, P., Kahn, C. R., Ogier-Denis, E., & Szanto, I. (2007). Reduced expression of the NADPH oxidase NOX4 is a hallmark of adipocyte differentiation. *Biochimica et Biophysica Acta, 1773* (7), 1015–1027. https://doi.org/10.1016/j.bbamcr.2007.03.003.

Munekata, K., & Sakamoto, K. (2009). Forkhead transcription factor Foxo1 is essential for adipocyte differentiation. *In Vitro Cellular & Developmental Biology Animal, 45*(10), 642–651. https://doi.org/10.1007/s11626-009-9230-5.

Murphy, S., Martin, S., & Parton, R. G. (2009). Lipid droplet-organelle interactions; sharing the fats. *Biochimica et Biophysica Acta, 1791*(6), 441–447. https://doi.org/10.1016/j.bbalip.2008.07.004.

Nakae, J., Kitamura, T., Kitamura, Y., Biggs, W. H., 3rd, Arden, K. C., & Accili, D. (2003). The forkhead transcription factor Foxo1 regulates adipocyte differentiation. *Developmental Cell, 4*(1), 119–129.

Nam, W. S., Park, K. M., & Park, J. W. (2012). RNA interference targeting cytosolic NADP(+)-dependent isocitrate dehydrogenase exerts anti-obesity effect in vitro and in vivo. *Biochimica et Biophysica Acta, 1822*(8), 1181–1188. https://doi.org/10.1016/j.bbadis.2012.04.003.

Nguyen, A., Guo, J., Banyard, D. A., Fadavi, D., Toranto, J. D., Wirth, G. A., Paydar, K. Z., Evans, G. R., & Widgerow, A. D. (2016). Stromal vascular fraction: A regenerative reality? Part 1: Current concepts and review of the literature. *Journal of Plastic, Reconstructive & Aesthetic Surgery: JPRAS, 69*(2), 170–179. https://doi.org/10.1016/j.bjps.2015.10.015.

Otto, T. C., & Lane, M. D. (2005). Adipose development: From stem cell to adipocyte. *Critical Reviews in Biochemistry and Molecular Biology, 40*(4), 229–242. https://doi.org/10.1080/10409230591008189.

Panday, A., Sahoo, M. K., Osorio, D., & Batra, S. (2015). NADPH oxidases: An overview from structure to innate immunity-associated pathologies. *Cellular & Molecular Immunology, 12*(1), 5–23. https://doi.org/10.1038/cmi.2014.89.

Pieralisi, A., Martini, C., Soto, D., Vila, M. C., Calvo, J. C., & Guerra, L. N. (2016). N-acetylcysteine inhibits lipid accumulation in mouse embryonic adipocytes. *Redox Biology, 9*, 39–44. https://doi.org/10.1016/j.redox.2016.05.006.

Quinlan, C. L., Orr, A. L., Perevoshchikova, I. V., Treberg, J. R., Ackrell, B. A., & Brand, M. D. (2012). Mitochondrial complex II can generate reactive oxygen species at high rates in both the forward and reverse reactions. *The Journal of Biological Chemistry, 287*(32), 27255–27264. https://doi.org/10.1074/jbc.M112.374629.

Riordan, N. H., Ichim, T. E., Min, W. P., Wang, H., Solano, F., Lara, F., Alfaro, M., Rodriguez, J. P., Harman, R. J., Patel, A. N., Murphy, M. P., Lee, R. R., & Minev, B. (2009). Non-expanded adipose stromal vascular fraction cell therapy for multiple sclerosis. *Journal of Translational Medicine, 7*, 29. https://doi.org/10.1186/1479-5876-7-29.

Rosen, E. D., & MacDougald, O. A. (2006). Adipocyte differentiation from the inside out. *Nature Reviews Molecular Cell Biology, 7*(12), 885–896. https://doi.org/10.1038/nrm2066.

Rossi, F., & Zatti, M. (1964). Biochemical aspects of phagocytosis in polymorphonuclear leucocytes. NADH and NADPH oxidation by the granules of resting and phagocytizing cells. *Experientia, 20*(1), 21–23.

Roy, J., Galano, J.-M., Durand, T., Le Guennec, J.-Y., & Lee, J. C.-Y. (2017). Physiological role of reactive oxygen species as promoters of natural defenses. *The FASEB Journal, 31*, 3729. https://doi.org/10.1096/fj.201700170R.

Ryu, J. M., Lee, H. J., Jung, Y. H., Lee, K. H., Kim, D. I., Kim, J. Y., Ko, S. H., Choi, G. E., Chai, I. I., Song, E. J., JY, O., Lee, S. J., & Han, H. J. (2015). Regulation of stem cell fate by ROS-mediated alteration of metabolism. *International Journal of Stem Cells, 8*(1), 24–35. 10.15283/ijsc.2015.8.1.24.

Sabharwal, S. S., & Schumacker, P. T. (2014). Mitochondrial ROS in cancer: Initiators, amplifiers or an Achilles' heel? *Nature Reviews Cancer, 14*(11), 709–721. https://doi.org/10.1038/nrc3803.

Saitoh, Y., Xiao, L., Mizuno, H., Kato, S., Aoshima, H., Taira, H., Kokubo, K., & Miwa, N. (2010). Novel polyhydroxylated fullerene suppresses intracellular oxidative stress together with repression of intracellular lipid accumulation during the differentiation of OP9 preadipocytes into adipocytes. *Free Radical Research, 44*(9), 1072–1081. https://doi.org/10.3109/10715762.2010.499905.

Samuelsson, L., Stromberg, K., Vikman, K., Bjursell, G., & Enerback, S. (1991). The CCAAT/enhancer binding protein and its role in adipocyte differentiation: Evidence for direct involvement in terminal adipocyte development. *The EMBO Journal, 10*(12), 3787–3793.

Samuni, Y., Cook, J. A., Choudhuri, R., Degraff, W., Sowers, A. L., Krishna, M. C., & Mitchell, J. B. (2010). Inhibition of adipogenesis by Tempol in 3T3-L1 cells. *Free Radical Biology & Medicine, 49*(4), 667–673. https://doi.org/10.1016/j.freeradbiomed.2010.05.028.

Satish, L., Krill-Burger, J. M., Gallo, P. H., Etages, S. D., Liu, F., Philips, B. J., Ravuri, S., Marra, K. G., LaFramboise, W. A., Kathju, S., & Rubin, J. P. (2015). Expression analysis of human adipose-derived stem cells during in vitro differentiation to an adipocyte lineage. *BMC Medical Genomics, 8*, 41. https://doi.org/10.1186/s12920-015-0119-8.

Sbarra, A. J., & Karnovsky, M. L. (1959). The biochemical basis of phagocytosis. I. Metabolic changes during the ingestion of particles by polymorphonuclear leukocytes. *The Journal of Biological Chemistry, 234*(6), 1355–1362.

Schroder, K., Wandzioch, K., Helmcke, I., & Brandes, R. P. (2009). Nox4 acts as a switch between differentiation and proliferation in preadipocytes. *Arteriosclerosis, Thrombosis, and Vascular Biology, 29*(2), 239–245. https://doi.org/10.1161/atvbaha.108.174219.

Slauch, J. M. (2011). How does the oxidative burst of macrophages kill bacteria? Still an open question. *Molecular Microbiology, 80*(3), 580–583. https://doi.org/10.1111/j.1365-2958.2011.07612.x.

Stephens, J. M. (2012). The fat controller: Adipocyte development. *PLoS Biology, 10*(11), e1001436. https://doi.org/10.1371/journal.pbio.1001436.

Tahara, E. B., Navarete, F. D., & Kowaltowski, A. J. (2009). Tissue-, substrate-, and site-specific characteristics of mitochondrial reactive oxygen species generation. *Free Radical Biology & Medicine, 46*(9), 1283–1297. https://doi.org/10.1016/j.freeradbiomed.2009.02.008.

Thomas, D., & Apovian, C. (2017). Macrophage functions in lean and obese adipose tissue. *Metabolism: Clinical and Experimental, 72*, 120–143. https://doi.org/10.1016/j.metabol.2017.04.005.

Tormos, K. V., Anso, E., Hamanaka, R. B., Eisenbart, J., Joseph, J., Kalyanaraman, B., & Chandel, N. S. (2011). Mitochondrial complex III ROS regulate adipocyte differentiation. *Cell Metabolism, 14*(4), 537–544. https://doi.org/10.1016/j.cmet.2011.08.007.

Turrens, J. F. (2003). Mitochondrial formation of reactive oxygen species. *The Journal of Physiology, 552*(Pt 2), 335–344. https://doi.org/10.1113/jphysiol.2003.049478.

Vigilanza, P., Aquilano, K., Baldelli, S., Rotilio, G., & Ciriolo, M. R. (2011). Modulation of intracellular glutathione affects adipogenesis in 3T3-L1 cells. *Journal of Cellular Physiology, 226*(8), 2016–2024. https://doi.org/10.1002/jcp.22542.

Virtue, S., & Vidal-Puig, A. (2010). Adipose tissue expandability, lipotoxicity and the metabolic syndrome--an allostatic perspective. *Biochimica et Biophysica Acta, 1801*(3), 338–349. https://doi.org/10.1016/j.bbalip.2009.12.006.

Wang, Y., Hudak, C., & Sul, H. S. (2010a). Role of preadipocyte factor 1 in adipocyte differentiation. *Clinical Lipidology, 5*(1), 109–115. https://doi.org/10.2217/clp.09.80.

Wang, Y., Zhao, L., Smas, C., & Sul, H. S. (2010b). Pref-1 interacts with fibronectin to inhibit adipocyte differentiation. *Molecular and Cellular Biology, 30*(14), 3480–3492. https://doi.org/10.1128/mcb.00057-10.

Wang, W., Zhang, Y., Lu, W., & Liu, K. (2015). Mitochondrial reactive oxygen species regulate adipocyte differentiation of mesenchymal stem cells in hematopoietic stress induced by arabinosylcytosine. *PLoS One, 10*(3), e0120629. https://doi.org/10.1371/journal.pone.0120629.

Weisberg, S. P., McCann, D., Desai, M., Rosenbaum, M., Leibel, R. L., & Ferrante, A. W., Jr. (2003). Obesity is associated with macrophage accumulation in adipose tissue. *The Journal of Clinical Investigation, 112*(12), 1796–1808. https://doi.org/10.1172/jci19246.

Yan, H., Aziz, E., Shillabeer, G., Wong, A., Shanghavi, D., Kermouni, A., Abdel-Hafez, M., & Lau, D. C. (2002). Nitric oxide promotes differentiation of rat white preadipocytes in culture. *Journal of Lipid Research, 43*(12), 2123–2129.

Yang, J. Y., Della-Fera, M. A., Rayalam, S., Ambati, S., Hartzell, D. L., Park, H. J., & Baile, C. A. (2008). Enhanced inhibition of adipogenesis and induction of apoptosis in 3T3-L1 adipocytes with combinations of resveratrol and quercetin. *Life Sciences, 82*(19–20), 1032–1039. https://doi.org/10.1016/j.lfs.2008.03.003.

Yoshihara, A., Hara, T., Kawashima, A., Akama, T., Tanigawa, K., Wu, H., Sue, M., Ishido, Y., Hiroi, N., Ishii, N., Yoshino, G., & Suzuki, K. (2012). Regulation of dual oxidase expression and H2O2 production by thyroglobulin. *Thyroid : Official Journal of the American Thyroid Association, 22*(10), 1054–1062. https://doi.org/10.1089/thy.2012.0003.

Younce, C., & Kolattukudy, P. (2012). MCP-1 induced protein promotes adipogenesis via oxidative stress, endoplasmic reticulum stress and autophagy. *Cellular Physiology and Biochemistry : International Journal of Experimental Cellular Physiology, Biochemistry, and Pharmacology, 30*(2), 307–320. https://doi.org/10.1159/000339066.

Zhang, M., Ikeda, K., JW, X., Yamori, Y., Gao, X. M., & Zhang, B. L. (2009). Genistein suppresses adipogenesis of 3T3-L1 cells via multiple signal pathways. *Phytotherapy Research: PTR, 23*(5), 713–718. https://doi.org/10.1002/ptr.2724.

Zhao, W.-E., Fan, J., Gao, R., & Ngoc, N. B. (2017). Suppressive effects of carotenoids on proliferation and differentiation of 3T3-L1 Preadipocytes. *Journal of Food and Nutrition Research, 5*(2), 129–136. 10.12691/jfnr-5-2-9.

Advs Exp. Medicine, Biology - Innovations in Cancer Research and Regenerative Medicine (2018) 1: 145–156
DOI 10.1007/5584_2017_127

Published online: 9 February 2018

Intravenous Infusion of Human Adipose Tissue-Derived Mesenchymal Stem Cells to Treat Type 1 Diabetic Mellitus in Mice: An Evaluation of Grafted Cell Doses

Loan Thi-Tung Dang, Anh Nguyen-Tu Bui, Cong Le-Thanh Nguyen, Nhat Chau Truong, Anh Thi-Van Bui, Ngoc Phan Kim, Kiet Dinh Truong, and Phuc Van Pham

Abstract

Mesenchymal stem cell (MSC) transplantation is a novel treatment for diabetes mellitus, especially type 1 diabetes. Many recent publications have demonstrated the efficacy of MSC transplantation on reducing blood glucose and increasing insulin production in both preclinical and clinical trials. However, the investigation of grafted cell doses has been lacking. Therefore, this study aimed to evaluate the different doses of MSCs on treatment of type 1 diabetes in mouse models. MSCs were isolated and expanded from human adipose tissue. Streptozotocin (STZ)-induced diabetic mice were divided into two groups that were intravenously transfused with two different doses of human MSCs: 10^6 or 2.10^6 cells/mouse. After transplantation, both grafted and placebo mice were monitored weekly for their blood glucose levels, glucose and insulin tolerance, pancreatic structural changes, and insulin production for 56 days after transplantation. The results showed that the higher dose of MSCs (2.10^6 cells/mouse) remarkably reduced death rate. The death rates were 50%, 66%, and 0% in placebo group, low-dose (1.10^6 MSCs) group, and high-dose (2.10^6 MSCs) group, respectively, after 56 days of treatment. Moreover, blood glucose levels were lower for the high-dose group compared to other groups. Glucose and insulin tolerance, as well as insulin production, were significantly improved in mice transplanted with 2.10^6 cells. The histochemical analyses also support these results. Thus, a higher (e.g., 2.10^6) dose of MSCs may be an effective dose for treatment of type 1 diabetes mellitus.

Keywords

Diabetes mellitus · Stem cells · Adipose-derived stem cells · Mesenchymal stem cells · Cell dose · Islet regeneration

L. T.-T. Dang, A. N.-T. Bui, and N. P. Kim
Stem Cell Institute, University of Science, VNUHCM, Ho Chi Minh city, Vietnam

C. Le-Thanh Nguyen, N. C. Truong, A. T.-V. Bui, and P. Van Pham (✉)
Stem Cell Institute, University of Science, VNUHCM, Ho Chi Minh city, Vietnam

Laboratory of Stem Cell Research and Application, University of Science, VNUHCM, Ho Chi Minh city, Vietnam
e-mail: pvphuc@hcmuns.edu.vn

K. D. Truong
Medical Genetics Institute, Ho Chi Minh city, Vietnam

Abbreviations

hADSCs	Human adipose-derived stem cells
GTT	Glucose tolerance test
H&E	Hematoxylin and eosin
MSCs	Mesenchymal stem cells
STZ	Streptozotocin
Th1/Th2	T helper 1/T helper 2
TNF-α	Tumor necrosis factor-alpha
IL-10	Interleukin-10
IL-12	Interleukin-12
IFN-γ	Interferon-gamma
EGF	Epidermal growth factor
bFGF	Basic fibroblast growth factor
PDGF	Platelet-derived growth factor
TGF-β	Transforming growth factor-beta
VEGF	Vascular endothelial growth factor
HGF	Hepatocyte growth factor
IGF	Insulin growth factor-1
NO	Nitric oxide
PGE2	Prostaglandin E2
IDO	Indoleamine 2,3-dioxygenase

1 Introduction

Diabetes is considered as an epidemic disease with a rapidly increasing incidence rate. Approximately 382 million people globally are affected by diabetes; the rate is expected to increase up to 592 million by 2035 (Guariguata et al. 2014). Recently, there have been many new advances in the treatment of diabetes such that the incidence has been managed and disease progression has been slowed.

Stem cell therapy has been proven to be a promising therapy for diabetes treatment according to numerous studies (Aghazadeh and Nostro 2017; Bhansali et al. 2017; Sood et al. 2017; Wehbe et al. 2016). Various kinds of stem cells have been evaluated for the treatment of both diabetes mellitus type 1 and 2 in preclinical and clinical trials. These stem cells have included hematopoietic stem cells (HSCs) (Cantu-Rodriguez et al. 2016; Li et al. 2012; Ye et al. 2017), mesenchymal stem cells (MSCs) (Mohammadi Ayenehdeh et al. 2017; Rahavi et al. 2015) (Ho et al. 2012; Hu et al. 2015; Kono et al. 2014; Lee et al. 2006; Sordi et al. 2010; Tsai et al. 2015), as well as others. According to finding from these studies, MSC transplantation offers several advantages for application to clinical treatment, namely, the relative ease of in vitro MSC expansion, immunogenicity of MSCs, and relative low cost of production.

Unlike HSCs, MSCs can improve the diabetes mellitus by various mechanisms. Most studies have suggested that MSCs can modulate the host immune system, leading to reduced inflammation and insulin tolerance. MSCs also can produce paracrine factors which trigger local stem cells to undergo self-renewal and differentiation to functional cells; these paracrine factors are also important for stimulating the healing of injured cells. Finally, a few studies have also demonstrated that MSCs can home to injured sites and differentiate into functional beta cells to regenerate the pancreatic tissues.

Although many studies have shown that MSC transplantation has benefits for type 1 diabetes mellitus, there are some studies that have shown limited results with low improvement of diabetes mellitus symptoms. Indeed, the outcome of MSC transplantation can be affected by various factors, such as cell source and cell properties, cell manipulation techniques, and transplanted strategies (e.g., cell doses/frequency, cell modification, and transplanted routes) (Gao et al. 2016).

This study aimed to investigate the optimal dose of MSCs for improving the diabetic condition in mice. MSCs from adipose tissue were transplanted into STZ-induced diabetic mice at two different doses: 1.10^6 and 2.10^6 cells/mouse, via intravenous route, to determine the safest and most effective dose for treatment of diabetes.

2 Materials and Methods

2.1 Stem Cell Preparation

Human adipose tissues were obtained from liposuction surgery with informed consent. All procedures in this study were approved by the local ethics committee. Adipose-derived stem cells (ADSCs) were cultured and identified following the established procedures described previously (Pham et al. 2014). Briefly, the stromal vascular fraction (SVF) from adipose tissue was collected using the Cell Extraction Kit (RegenmedLab, Vietnam) according to the manufacturer's instructions. The SVFs were cultured in MSCCult medium (DMEM/F12 supplemented with EGF, FGF). MSC candidates at the third passage were confirmed as MSCs using criteria such as shape as evaluated under inverted phase-contrast microscopy, surface marker expression (CD14, CD34, CD44, CD73, CD166, etc.), and the adipogenic and osteogenic differentiation potential. Flow cytometry as well as in vitro differentiation procedures were similar to previously described (Pham et al. 2014).

2.2 STZ-Induced Diabetic Mice

All animal procedures were in compliance with the instructions and principles of the Animal Care and Use Committee. Eight-week-old male mice with 25 $\pm$ 2.5 g were housed in 12 h light/12 h dark cycle and fasted overnight before STZ administration. To induce type 1 diabetes mellitus, the mice were injected with a single intraperitoneal dose of 100 mg/kg STZ (Santa Cruz Biotechnology, Dallas, TX). Blood glucose was examined by OneTouch Ultra (LifeScan/Johnson & Johnson, New Brunswick, NJ) 3 days after STZ injection. The mice were considered to be diabetic if their blood glucose levels were stable at 300 mg/dl or above on three consecutive weekly readings. For glucose or insulin tolerance tests, mice were fasted 12 h and intraperitoneally injected with either glucose (2 g/kg) or insulin (0.75 mg/kg). The blood glucose concentrations were measured at 15, 30, 60, and 120 min thereafter.

2.3 Stem Cell Transplantation

ADSCs from the third to fifth passages were de-attached from plastic surfaces by trypsin/EDTA (0.25%) and then carefully washed twice with saline buffer. The cell suspension was filtered through a 70 μm membrane to obtain single cells. The cell suspension was diluted to achieve doses of 1.10^6 or 2.10^6 cells/200 μl/dose.

Eighteen diabetic mice were randomly divided into three groups: G2 (placebo group, i.e., saline infusion; n = 6), G3 (treatment group, i.e., intravenous infusion of 10^6 cells; n = 6), and G4 (treatment group, i.e., intravenous infusion of 2.10^6 cells; n = 6). All mice in the three groups were monitored for glucose metabolism and islet function for 8 weeks. The normal (nondiabetic) mice were labeled as group G1 (n = 6).

2.4 Measurement of Insulin Release

Mouse blood from facial vein was spun at 3000 rpm for 10 min at 4 °C to obtain serum; insulin levels were measured by using Ultrasensi tive Mouse Insulin ELISA (Mercodia, Sweden) according to the manufacturer's instructions. In brief, the serum and calibrators were added into the wells and the enzyme was added afterward. Each sample was measured repeatedly three times. The plate was incubated for 2 h and carefully washed before TMB substrate was added. The reactions were terminated by adding stop solution, and the contents were read within 30 min at OD 450 nm by DTX 880 Multimode Detector (Beckman Coulter, USA). The calibration curve was constructed, and the insulin concentrations were interpolated from the standard curve using Prism 6 (GraphPad Software, San Diego, CA).

2.5 Histological and Immunohistochemistry Analyses

For histological examination, mouse pancreas was fixed in formalin and embedded in paraffin blocks. The tissues were cut into 10-μm-thin sections and stained with hematoxylin and eosin following established protocols. The samples were examined by microscopy; images were analyzed by AxioVision microscope software

(Carl Zeiss, Germany) to quantify islets and measure islet size. For immunofluorescence staining, mouse pancreas was fixed in 4% paraformaldehyde, dipped in sucrose, embedded in optimal cutting temperature (OCT) compound, and cut into 10-μm-thin sections. These slides were then dipped in 5% bovine serum albumin (BSA) overnight at 4 °C. The primary anti-insulin antibody (Santa Cruz Biotechnology, Dallas, TX) was added, and the samples were incubated overnight at 4 °C and then stained with FITC-conjugated secondary antibody (Sigma-Aldrich, St. Louis, MO) for 2 h. Finally, the nuclei were stained with 4′,6-diamidino-2-phenylindole (DAPI) (Santa Cruz Biotechnology) and the sections were observed under fluorescence microscopy (Cell Observer, Carl Zeiss, Germany).

2.6 Data Analysis

GraphPad Prism 6 (GraphPad Software, Inc.) was used to analyze the data. Statistical significance was defined as $P < 0.05$, and results were analyzed by student's unpaired t-test and one-way ANOVA. Data were reported as the mean and standard error of the mean (SEM).

3 Results

3.1 Human Adipose-Derived Stem Cells Express Specific MSC Phenotype

The results showed that the obtained ADSCs expressed the standard phenotype of MSCs. Indeed, they could adhere to surface of flasks with fibroblast-like shape (Fig. 1f). They also expressed the common markers of MSCs; they were positive for CD44, CD73, and CD166, while negative for CD14 and CD34 (Fig. 1a–e). In the induced medium, they were able to successfully differentiate into adipocytes and osteoblasts (Fig. 1g–i).

3.2 Higher Dose of ADSCs Prolongs the Survival of Diabetic Mice and Improves Weight Loss

Without any treatment, the diabetic mice die due to hyperglycemia and its complications, starting from day 21. Moreover, the dose of 10^6 MSCs (G3) did not improve survival of diabetic mice; 66% (4/6) of mice died in G3. However, 50% (3/6) of mice died in G2 (placebo) at day 56. On the other hand, 100% of mice (6/6) in G4 (2.10^6 MSCs) and 100% of those in G1 (normal mice) survived until the end of follow-up time. In general, MSC transplantation did not cause any acute responses or immediate death in mice. These data suggest that ADSC transplantation is safe for normal and STZ-induced diabetic mice and that a dose of 2.10^6 MSCs could enhance mouse survival (Fig. 2).

Moreover, high-dose infusion (G4; 2.10^6 MSCs) could improve weight loss in diabetic mice from day 7 (30.9 ± 1.9 g) to day 56 (39.3 ± 1.4 g), in comparison to untreated diabetic mice (25.2 ± 0.3 and 22.3 ± 0.3 g, respectively). There was no difference between the mice treated with the low-dose (G3, 10^6 MSCs) and placebo treatment (G2).

3.3 Transplantation of 2.10^6 ADSCs Improves Glucose Homeostasis of Diabetic Mice

The diabetic mice in G4 showed significant improvement of glycemia from day 21 to day 35, in comparison with groups G3 and G2 ($p < 0.05$) (Fig. 3a). The G2 and G3 mouse groups showed increased blood glucose levels (370 ± 24 and 409 ± 12 mg/dL at day 0 and 518 ± 30 and 518 ± 20 mg/dL at day 56, respectively). Meanwhile mice in the group G4 (high-dose group) showed a gradual decrease of glycemia (from 380 ± 25 mg/dL at day 0 to 306 ± 32 mg/dL at day 56) ($p < 0.05$). These results were supported by the plasma insulin levels measured at day 56 after treatment (Fig. 3b). The significant improvement of insulin

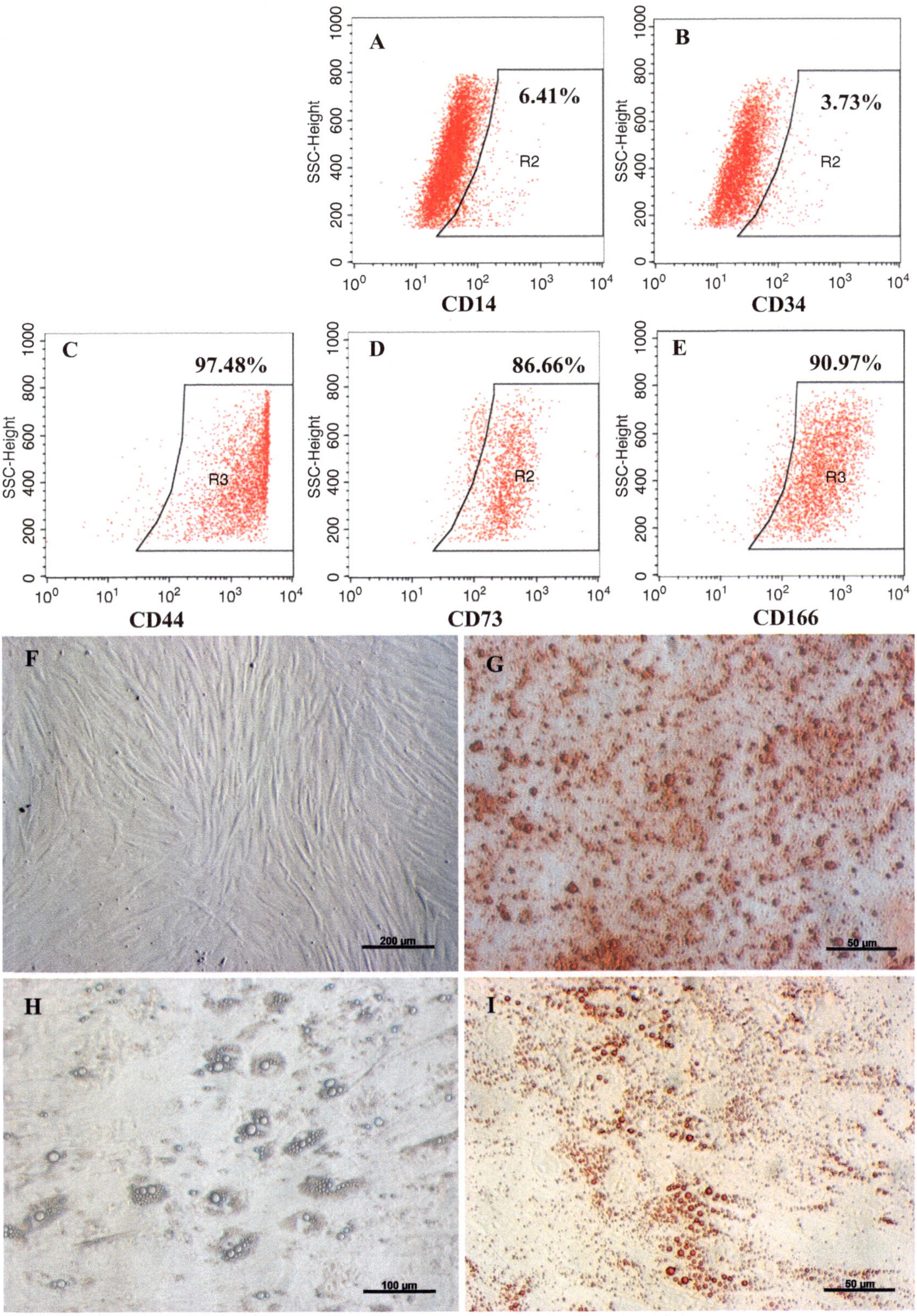

Fig. 1 **Characterization of human adipose-derived stem cells.** Surface markers as determined by flow cytometry (**a**–**e**); fibroblast-like morphology (100X, scale bar = 200 μm) (**f**) and ADSC differentiation into osteoblasts as stained by alizarin red (200X, scale bar = 100 μm) (**g**); ADSC differentiation into fat cells (lipid droplets) (**h**) and fat cells as stained by oil red (scale bar = 200 μm) (**i**)

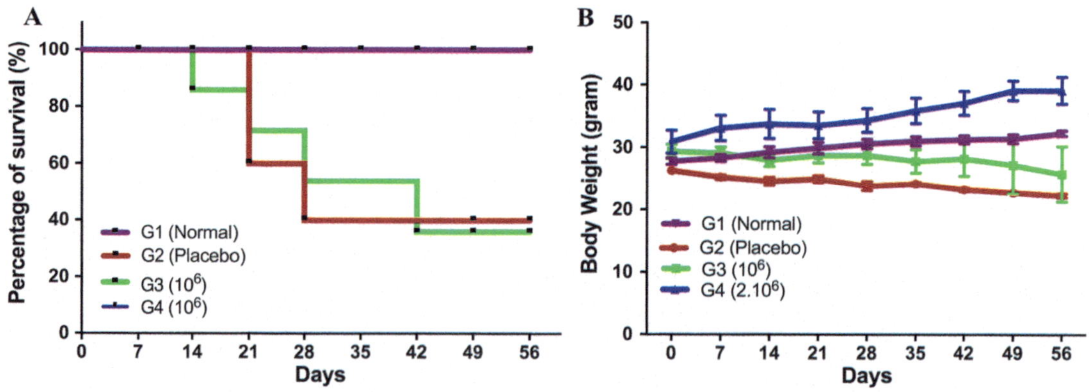

Fig. 2 Kaplan–Meier curves for survival rate and changes in body weight in some groups G1, G2, G3, and G4. The results showed that there were significant differences about survival and body weight improvement between mice in G3 and G4

secretion compared to before treatment was found in G4 (high-dose) mice, while the G2 and G3 diabetic mice had very low levels of plasma insulin. Moreover, the glucose tolerance test (GTT) (Fig. 3c) confirmed the enhanced function of beta cells in mice transplanted with the higher dose of ADSCs (G3). This was likely due to the fact that the blood glucose levels remarkably decreased 30 min after glucose injection and reached the initial value 120 min after glucose injection.

3.4 The Islets of ADSC-Treated Diabetic Mice Were Prevented from Destruction

At the end of follow-up time, the pancreatic tissues were collected and fixed before they were cut and stained with hematoxylin and eosin (H&E) or stained with anti-insulin antibody in immunohistochemistry analysis (Fig. 4). The H&E stained samples of pancreatic tissues showed different types of islet structure: normal islets had apparent organization comprised of relatively homomorphic cells and surrounded by enlarged acinar cells (Fig. 4a). These typical islets appeared predominantly in normal (G1) mice, while they were almost absent in the pancreas of G2 mice. On the contrary, in diabetic mice, the altered islets showed atrophied shape and were difficult to distinguish from acinar structures. Loss of islet cells and their uneven distribution were also observed in diabetic pancreas (Fig. 4b). Moreover, several islets were detected in which organization was intact although their shape was changed and size possibly reduced (Fig. 4c). These islets were found remarkably in G4 mice, of which blood glucose levels were improved after MSC transplantation. The differences of islets among the STZ-induced diabetic mice (G2, G3) and normal mice (G1), as well as recovered islets found in high-dose MSC-treated mice (G4), were confirmed by immunofluorescence analysis. In the islets of normal mice, the strong expression of insulin demonstrated the predominance of beta cells clustered in the center of the islets (Fig. 4d). Insulin signals were decreased according to the damage stages of diabetic islets found in G2, G3, and G4 mice (Fig. 4e, f).

Analysis of islet quantity and size revealed that STZ injection not only destroyed many islets but also reduced islet size (Fig. 5). Additionally, transplantation of ADSCs at a dose of 2.10^6 cells (G4) could enhance two parameters: islet quantity and mean islet size. The islet quantity was doubled in G4 mice (5.5 ± 0.3 islets/section) in comparison with the placebo group (G2) (2.2 ± 0.2 islets/section); the mean of islet size was 116 ± 6 μm in G4 mice compared with 72 ± 4 μm in G2 mice ($p < 0.001$). However, the dose of 10^6 cells did not improve the number of islets nor islet size.

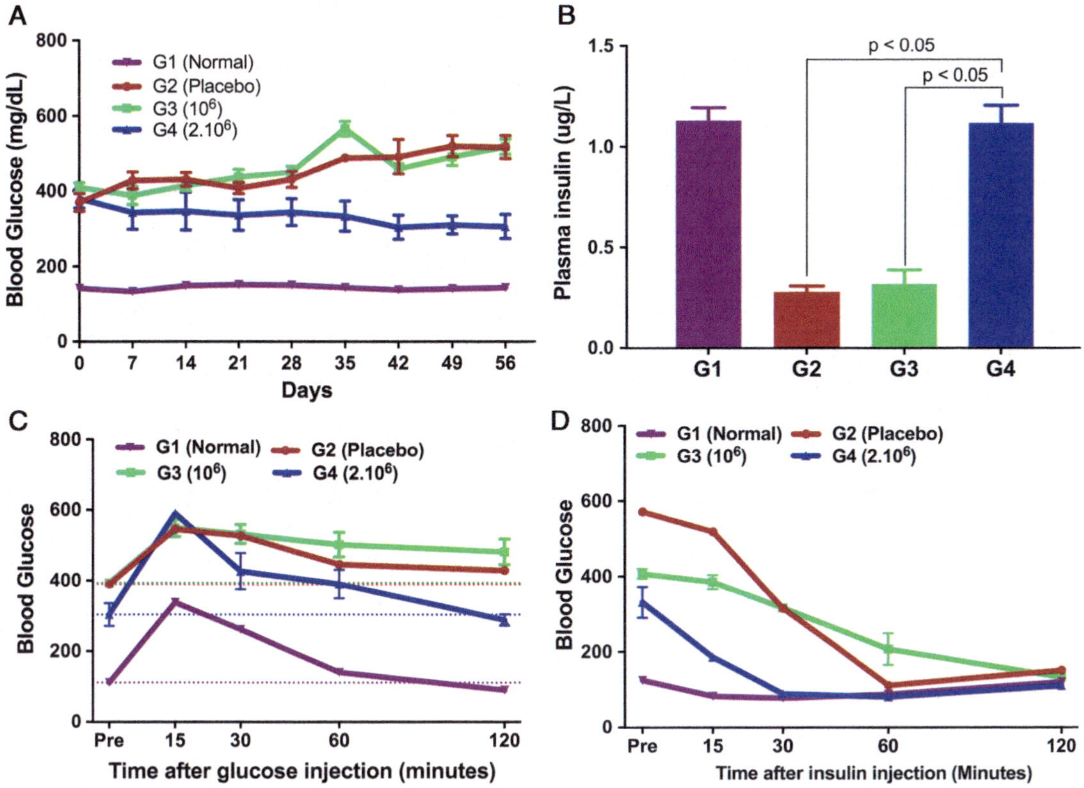

Fig. 3 Transplantation of 2.10^6 ADSCs improves hyperglycemia. (**a**) Blood glucose levels; (**b**) serum insulin levels; (**c**) glucose tolerance; and (**d**) insulin tolerance. Purple, nondiabetic (normal) mice (G1); red, diabetic mice receiving placebo (saline) (G2); green, diabetic mice receiving 10^6 MSC dose (G3); and blue, diabetic mice receiving 2.10^6 MSC dose (G4)

4 Discussion

Type 1 diabetes accounts for 5–10% of all diabetic cases in the world. Without treatment, type 1 diabetic patients may experience many serious complications and even death. Conventional methods such as insulin injection and islet transplantation are not optimal treatment strategies in terms of their side effects and limited supply. Meanwhile, MSC therapy has been shown to be a promising strategy because of the prominent therapeutic properties of MSCs, such as low expression of MHC class II and secretion of many cytokines responsible for modulating tissue environment, regeneration, and the healing process. More importantly, MSCs can be differentiated into insulin-producing cells (Gabr et al. 2013; Kadam and Bhonde 2010; Kao et al. 2015; Karnieli et al. 2007; Marappagounder et al. 2013; Moshtagh et al. 2013; Okura et al. 2009; Seyedi et al. 2016; Thi-Tung Dang et al. 2015; Pham et al. 2014). Therefore, this study used MSCs as a platform to treat diabetes mellitus type 1.

In the first experiment, MSCs were successfully isolated from human adipose tissues. These cells exhibit full characteristics of standard MSCs as suggested by Dominici et al. (2006), including adherence to flask surface with fibroblast-like shape, expression of particular MSC markers, and successful differentiation into mesoderm-derived types, such as adipocytes and osteoblasts. These cells were expanded to the fifth passage and used in subsequent experiments.

In the second experiment, we produced the type 1 diabetic mellitus mice by injection of a

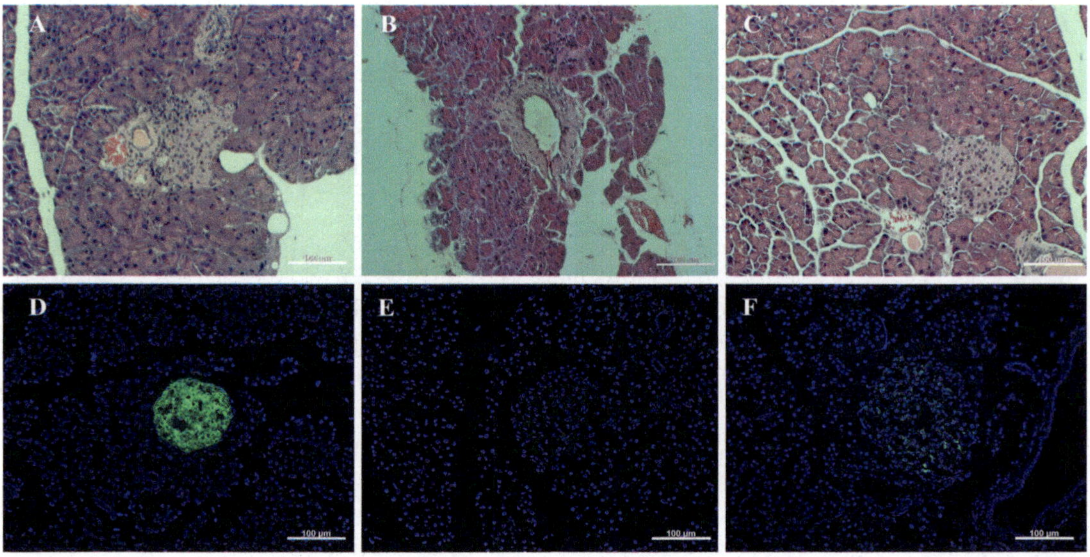

Fig. 4 Pancreatic sections stained with hematoxylin and eosin (a, b, c) and stained with immunofluorescent anti-insulin antibody (d, e, f). Typical structure of islet positive for insulin (**a**, **d**); STZ-induced diabetic islets had aberrant morphology and negative expression of insulin (**b**, **e**); and the diabetic islets prevented from destruction expressed poor insulin signal (**c**, **f**) (insulin expression as green signal and nuclei stained with Hoechst 33342 as blue signal)

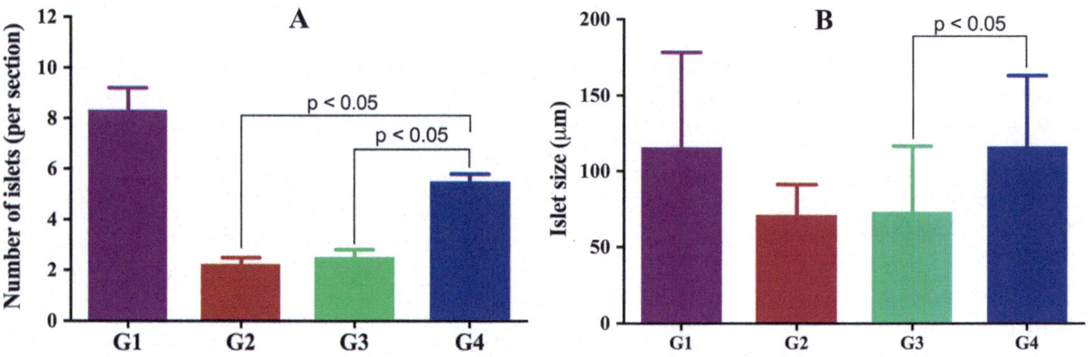

Fig. 5 The differences in islet size and number of islets among the treatment groups. Four slides in each sample were analyzed; data is expressed as means ± SEM (n = 3 samples for each group)

single high dose of streptozotocin (100 mg/kg), causing diabetes in animals due to destruction of pancreatic beta cells. Indeed, the STZ-induced model of diabetes resembles the pathophysiology of type 1 diabetes mellitus in humans. The diabetic pancreas induced by STZ showed loss of islets and islet size reduction. Moreover, the diabetic islets were transformed in shape and structure and showed decreased insulin secretion. Thus, STZ-treated mice (approximately threefold compared to normal/nondiabetic mice) showed hyperglycemia and impaired glucose tolerance. These mice were used to investigate the effects of ADSCs on diabetes using two distinct doses: 10^6 and 2.10^6 cells.

The results showed that intravenous injection of ADSCs was a safe treatment in mice since there were no adverse effects observed within the first 2 weeks after transplantation. At day 21, 2/6 mice died in the placebo group, while all mice were alive in the high MSC dose group (G4). This suggested that ADSC transplantation did not cause mortality in diabetic mice and that the lower MSC dose did not save mice from death by hyperglycemia complications. Meanwhile, the diabetic mice in G4 (treated with higher dose of cells) and normal mice (G1) survived until day 56 and thereafter (data not shown). Therefore, ADSC transplantation at the 2.10^6 MSC dose had a significant effect on enhancing vitality and survival of diabetic mice.

For 8 weeks after ADSC transplantation, the results showed that the dose of 10^6 cells did not ameliorate the diabetic condition. This cell dose seemed to be insufficient for producing reactions and impacts necessary to improve or cure the damaged islets. Meanwhile, the dose of 2.10^6 cells could improve diabetes in G4 mice owing to alleviation of hyperglycemia. Specifically, at day 56, blood glucose of G4 mice decreased by approximately 20% compared to day 0, while blood glucose increased by 40% in G2 mice. The recovered mice also showed amelioration of glucose tolerance and insulin production. These improvements in pancreatic function may result from islet quantity increasing and/or prevention of islets from damage. Indeed, ADSC transplantation contributed to reducing the loss of islet cells and increasing islet size. Therefore, the MSC-treated mice showed significantly improved health and survival.

Human MSC transplantation has been demonstrated to be an effective treatment in type 1 diabetic animals in some previous publications (Bell et al. 2012a; Bell et al. 2012b; Ho et al. 2012; Kono et al. 2014; Zhou et al. 2015). Although these human MSCs were detected at low count in mouse pancreases (Kono et al. 2014; Meyerrose et al. 2007; Zhou et al. 2015), it has not been clear whether the low counts arose from issues of homing, transdifferentiation, and/or replacement of MSCs (Zhou et al. 2016). Human MSCs clearly possess strong immunosuppressive effects (Atoui and Chiu 2012) and are known to produce many kinds of cytokines and signaling factors which affect not only immune responses but tissue environment and cell growth.

Moreover, the hypo-immunogenic properties of xenografted MSCs enable them to survive long enough to induce modifications in the pancreatic environment without any immunosuppressant treatment. In fact, MSCs have been demonstrated to secrete immunomodulatory factors which change the balance between immune cells, in terms of cell–cell contact and humoral activities. MSC-mediated inhibition affects natural killer cells, dendritic cells, and several subtypes of B and T lymphocytes via the production of MSC-derived agents (e.g., IDO, NO, PGE2, TGF-β, TNF-α, IL-10, IL-12, and IFN-γ) (Gao et al. 2016; Ma and Chan 2016; Spaggiari et al. 2008; Tsai et al. 2015; Yaochite et al. 2016). Moreover, MSCs can improve the proliferation of regulatory T cells and restore the Th1/Th2 balance leading to anti-inflammation (Ezquer et al. 2012); thus, its effects may result in decreased cell death. Most importantly, some recently published data have revealed that MSCs produce many trophic factors (e.g., IGF, EGF, PDGF, HGF, and bFGF). VEGF is associated with the modification of the pancreatic environment and enhancement of local endogenous stem cell proliferation as well as vascular regeneration via paracrine effects (Nagaishi et al. 2016; Youssef et al. 2017; Zhou et al. 2015, 2016). These effects could prevent pancreatic islets from destruction and/or enhance islet regeneration and function, thus ameliorating diabetes.

Overall, in this study, we found that MSC transplantation at the higher dose (2.10^6 cells), as compared to lower dose (10^6), was consistently more effective for alleviating type 1 diabetes. It is likely that the dose of 2.10^6 cells is the "threshold" to trigger many processes related to modulation, repair, and regeneration in pancreatic tissues. These effects may lead to amelioration of diabetes.

5 Conclusion

In conclusion, this study showed that xenotransplantation of human adipose-derived stem cells could be applied to type 1 diabetic mice. Moreover, the dose of 2.10^6 cells was shown to be more effective than 10^6 cells in terms of its effect of improving glucose metabolism and pancreatic function in diabetic mice. Type 1 diabetic mellitus mice transplanted with 2.10^6 ADSCs showed a remarkable improvement in islet structure, insulin secretion, blood glucose levels, and glucose tolerance. Consequently, ADSC-treated mice recovered their body weight as well as showing prolonged life. Our study provides experimental evidence of the effectiveness of MSCs for type 1 diabetes mellitus treatment.

Acknowledgment This research was funded by the Ministry of Science and Technology via project Grant No. -DTDL.2012-G/23 and Vietnam National University, Ho Chi Minh City, via project No. C2016-18-18.

Author Contribution LTTD, designed the study, performed the experiments, analyzed the data, and drafted the manuscript; PVP, KDT, designed the study and reviewed and corrected the manuscript; ANTB, CLTN, performed the experiments and analyzed the data and drafted the manuscript; NCT, ATVB, performed the experiments and reviewed and drafted the manuscript; and NPK analyzed the data and wrote the manuscript.

Competing Interests The authors declare that no competing interests exist.

References

Aghazadeh, Y., & Nostro, M. C. (2017). Cell therapy for type 1 diabetes: Current and future strategies. *Current Diabetes Reports, 17*, 37.

Atoui, R., & Chiu, R. C. (2012). Concise review: Immunomodulatory properties of mesenchymal stem cells in cellular transplantation: Update, controversies, and unknowns. *Stem Cells Translational Medicine, 1*, 200–205.

Bell, G. I., Broughton, H. C., Levac, K. D., Allan, D. A., Xenocostas, A., & Hess, D. A. (2012a). Transplanted human bone marrow progenitor subtypes stimulate endogenous islet regeneration and revascularization. *Stem Cells and Development, 21*, 97–109.

Bell, G. I., Meschino, M. T., Hughes-Large, J. M., Broughton, H. C., Xenocostas, A., & Hess, D. A. (2012b). Combinatorial human progenitor cell transplantation optimizes islet regeneration through secretion of paracrine factors. *Stem Cells and Development, 21*, 1863–1876.

Bhansali, S., Dutta, P., Kumar, V., Yadav, M. K., Jain, A., Mudaliar, S., Bhansali, S., Sharma, R. R., Jha, V., Marwaha, N., et al. (2017). Efficacy of autologous bone marrow-derived mesenchymal stem cell and mononuclear cell transplantation in type 2 diabetes mellitus: A randomized, placebo-controlled comparative study. *Stem Cells and Development, 26*, 471–481.

Cantu-Rodriguez, O. G., Lavalle-Gonzalez, F., Herrera-Rojas, M. A., Jaime-Perez, J. C., Hawing-Zarate, J. A., Gutierrez-Aguirre, C. H., Mancias-Guerra, C., Gonzalez-Llano, O., Zapata-Garrido, A., Villarreal-Perez, J. Z., et al. (2016). Long-term insulin independence in type 1 diabetes mellitus using a simplified autologous stem cell transplant. *The Journal of Clinical Endocrinology and Metabolism, 101*, 2141–2148.

Dominici, M., Le Blanc, K., Mueller, I., Slaper-Cortenbach, I., Marini, F., Krause, D., Deans, R., Keating, A., Prockop, D., & Horwitz, E. (2006). Minimal criteria for defining multipotent mesenchymal stromal cells. The International Society for Cellular Therapy position statement. *Cytotherapy, 8*, 315–317.

Ezquer, F., Ezquer, M., Contador, D., Ricca, M., Simon, V., & Conget, P. (2012). The antidiabetic effect of mesenchymal stem cells is unrelated to their transdifferentiation potential but to their capability to restore Th1/Th2 balance and to modify the pancreatic microenvironment. *Stem Cells, 30*, 1664–1674.

Gabr, M. M., Zakaria, M. M., Refaie, A. F., Ismail, A. M., Abou-El-Mahasen, M. A., Ashamallah, S. A., Khater, S. M., El-Halawani, S. M., Ibrahim, R. Y., Uin, G. S., et al. (2013). Insulin-producing cells from adult human bone marrow mesenchymal stem cells control streptozotocin-induced diabetes in nude mice. *Cell Transplantation, 22*, 133–145.

Gao, F., Chiu, S. M., Motan, D. A., Zhang, Z., Chen, L., Ji, H. L., Tse, H. F., Fu, Q. L., & Lian, Q. (2016). Mesenchymal stem cells and immunomodulation: Current status and future prospects. *Cell Death & Disease, 7*, e2062.

Guariguata, L., Whiting, D. R., Hambleton, I., Beagley, J., Linnenkamp, U., & Shaw, J. E. (2014). Global estimates of diabetes prevalence for 2013 and projections for 2035. *Diabetes Research and Clinical Practice, 103*, 137–149.

Ho, J. H., Tseng, T. C., Ma, W. H., Ong, W. K., Chen, Y. F., Chen, M. H., Lin, M. W., Hong, C. Y., & Lee, O. K. (2012). Multiple intravenous transplantations of mesenchymal stem cells effectively restore long-term blood glucose homeostasis by hepatic engraftment and beta-cell differentiation in streptozocin-induced diabetic mice. *Cell Transplantation, 21*, 997–1009.

Hu, J., Wang, Y., Wang, F., Wang, L., Yu, X., Sun, R., Wang, Z., Wang, L., Gao, H., Fu, Z., et al. (2015). Effect and mechanisms of human Wharton's jelly-derived mesenchymal stem cells on type 1 diabetes in NOD model. *Endocrine, 48*, 124–134.

Kadam, S. S., & Bhonde, R. R. (2010). Islet neogenesis from the constitutively nestin expressing human umbilical cord matrix derived mesenchymal stem cells. *Islets, 2*, 112–120.

Kao, S. Y., Shyu, J. F., Wang, H. S., Lin, C. H., Su, C. H., Chen, T. H., Weng, Z. C., & Tsai, P. J. (2015). Comparisons of differentiation potential in human mesenchymal stem cells from Wharton's jelly, bone marrow, and pancreatic tissues. *Stem Cells International, 2015*, 306158.

Karnieli, O., Izhar-Prato, Y., Bulvik, S., & Efrat, S. (2007). Generation of insulin-producing cells from human bone marrow mesenchymal stem cells by genetic manipulation. *Stem Cells, 25*, 2837–2844.

Kono, T. M., Sims, E. K., Moss, D. R., Yamamoto, W., Ahn, G., Diamond, J., Tong, X., Day, K. H., Territo, P. R., Hanenberg, H., et al. (2014). Human adipose-derived stromal/stem cells protect against STZ-induced hyperglycemia: Analysis of hASC-derived paracrine effectors. *Stem Cells, 32*, 1831–1842.

Lee, R. H., Seo, M. J., Reger, R. L., Spees, J. L., Pulin, A. A., Olson, S. D., & Prockop, D. J. (2006). Multipotent stromal cells from human marrow home to and promote repair of pancreatic islets and renal glomeruli in diabetic NOD/scid mice. *Proceedings of the National Academy of Sciences of the United States of America, 103*, 17438–17443.

Li, L., Shen, S., Ouyang, J., Hu, Y., Hu, L., Cui, W., Zhang, N., Zhuge, Y. Z., Chen, B., Xu, J., et al. (2012). Autologous hematopoietic stem cell transplantation modulates immunocompetent cells and improves beta-cell function in Chinese patients with new onset of type 1 diabetes. *The Journal of Clinical Endocrinology and Metabolism, 97*, 1729–1736.

Ma, O. K., & Chan, K. H. (2016). Immunomodulation by mesenchymal stem cells: Interplay between mesenchymal stem cells and regulatory lymphocytes. *World Journal of Stem Cells, 8*, 268–278.

Marappagounder, D., Somasundaram, I., Dorairaj, S., & Sankaran, R. J. (2013). Differentiation of mesenchymal stem cells derived from human bone marrow and subcutaneous adipose tissue into pancreatic islet-like clusters in vitro. *Cellular & Molecular Biology Letters, 18*, 75–88.

Meyerrose, T. E., De Ugarte, D. A., Hofling, A. A., Herrbrich, P. E., Cordonnier, T. D., Shultz, L. D., Eagon, J. C., Wirthlin, L., Sands, M. S., Hedrick, M. A., et al. (2007). In vivo distribution of human adipose-derived mesenchymal stem cells in novel xenotransplantation models. *Stem Cells, 25*, 220–227.

Mohammadi Ayenehdeh, J., Niknam, B., Rasouli, S., Hashemi, S. M., Rahavi, H., Rezaei, N., Soleimani, M., Liaeiha, A., Niknam, M. H., & Tajik, N. (2017). Immunomodulatory and protective effects of adipose tissue-derived mesenchymal stem cells in an allograft islet composite transplantation for experimental autoimmune type 1 diabetes. *Immunology Letters, 188*, 21–31.

Moshtagh, P. R., Emami, S. H., & Sharifi, A. M. (2013). Differentiation of human adipose-derived mesenchymal stem cell into insulin-producing cells: An in vitro study. *Journal of Physiology and Biochemistry, 69*, 451–458.

Nagaishi, K., Mizue, Y., Chikenji, T., Otani, M., Nakano, M., Konari, N., & Fujimiya, M. (2016). Mesenchymal stem cell therapy ameliorates diabetic nephropathy via the paracrine effect of renal trophic factors including exosomes. *Scientific Reports, 6*, 34842.

Okura, H., Komoda, H., Fumimoto, Y., Lee, C. M., Nishida, T., Sawa, Y., & Matsuyama, A. (2009). Transdifferentiation of human adipose tissue-derived stromal cells into insulin-producing clusters. *Journal of Artificial Organs, 12*, 123–130.

Pham, V. P., Vu, B. N., Phan, L. C. N., Le, M. D., Truong, C. N., Truong, H. N., Bui, H. T. K., & Phan, K. N. (2014). Good manufacturing practice-compliant isolation and culture of human adipose-derived stem cells. *Biomedical Research and Therapy, 1*, 133–141.

Rahavi, H., Hashemi, S. M., Soleimani, M., Mohammadi, J., & Tajik, N. (2015). Adipose tissue-derived mesenchymal stem cells exert in vitro immunomodulatory and beta cell protective functions in streptozotocin-induced diabetic mice model. *Journal of Diabetes Research, 2015*, 878535.

Seyedi, F., Farsinejad, A., Nematollahi-Mahani, S. A., Eslaminejad, T., & Nematollahi-Mahani, S. N. (2016). Suspension culture alters insulin secretion in induced human umbilical cord matrix-derived mesenchymal cells. *Cell Journal, 18*, 52–61.

Sood, V., Bhansali, A., Mittal, B. R., Singh, B., Marwaha, N., Jain, A., & Khandelwal, N. (2017). Autologous bone marrow derived stem cell therapy in patients with type 2 diabetes mellitus – defining adequate administration methods. *World Journal of Diabetes, 8*, 381–389.

Sordi, V., Melzi, R., Mercalli, A., Formicola, R., Doglioni, C., Tiboni, F., Ferrari, G., Nano, R., Chwalek, K., Lammert, E., et al. (2010). Mesenchymal cells appearing in pancreatic tissue culture are bone marrow-derived stem cells with the capacity to improve transplanted islet function. *Stem Cells, 28*, 140–151.

Spaggiari, G. M., Capobianco, A., Abdelrazik, H., Becchetti, F., Mingari, M. C., & Moretta, L. (2008). Mesenchymal stem cells inhibit natural killer-cell proliferation, cytotoxicity, and cytokine production: Role of indoleamine 2,3-dioxygenase and prostaglandin E2. *Blood, 111*, 1327–1333.

Thi-Tung Dang, L., Nguyen-Tu Bui, A., Minh Pham, V., Kim Phan, N., & Van Pham, P. (2015). Production of islet-like insulin-producing cell clusters in vitro from adipose derived stem cells. *Biomedical Research and Therapy, 2*, 184–192.

Tsai, P. J., Wang, H. S., Lin, G. J., Chou, S. C., Chu, T. H., Chuan, W. T., Lu, Y. J., Weng, Z. C., Su, C. H., Hsieh, P. S., et al. (2015). Undifferentiated Wharton's jelly mesenchymal stem cell transplantation induces insulin-producing cell differentiation and suppression of T-cell-mediated autoimmunity in nonobese diabetic mice. *Cell Transplantation, 24*, 1555–1570.

Van Pham, P., Thi-My Nguyen, P., Thai-Quynh Nguyen, A., Minh Pham, V., Nguyen-Tu Bui, A., Thi-Tung Dang, L., Gia Nguyen, K., & Kim Phan, N. (2014). Improved differentiation of umbilical cord blood-derived mesenchymal stem cells into insulin-producing cells by PDX-1 mRNA transfection. *Differentiation, 87*, 200–208.

Wehbe, T., Chahine, N. A., Sissi, S., Abou-Joaude, I., & Chalhoub, L. (2016). Bone marrow derived stem cell therapy for type 2 diabetes mellitus. *Stem Cell Investigation, 3*, 87.

Yaochite, J. N., de Lima, K. W., Caliari-Oliveira, C., Palma, P. V., Couri, C. E., Simoes, B. P., Covas, D. T., Voltarelli, J. C., Oliveira, M. C., Donadi, E. A., et al. (2016). Multipotent mesenchymal stromal cells from patients with newly diagnosed type 1 diabetes mellitus exhibit preserved in vitro and in vivo immunomodulatory properties. *Stem Cell Research & Therapy, 7*, 14.

Ye, L., Li, L., Wan, B., Yang, M., Hong, J., Gu, W., Wang, W., & Ning, G. (2017). Immune response after autologous hematopoietic stem cell transplantation in type 1 diabetes mellitus. *Stem Cell Research & Therapy, 8*, 90.

Youssef, A., Aboalola, D., & Han, V. K. (2017). The roles of insulin-like growth factors in mesenchymal stem cell niche. *Stem Cells International, 2017*, 9453108.

Zhou, Y., Hu, Q., Chen, F., Zhang, J., Guo, J., Wang, H., Gu, J., Ma, L., & Ho, G. (2015). Human umbilical cord matrix-derived stem cells exert trophic effects on beta-cell survival in diabetic rats and isolated islets. *Disease Models & Mechanisms, 8*, 1625–1633.

Zhou, J. Y., Zhang, Z., & Qian, G. S. (2016). Mesenchymal stem cells to treat diabetic neuropathy: A long and strenuous way from bench to the clinic. *Cell Death Discovery, 2*, 16055.

Advs Exp. Medicine, Biology - Innovations in Cancer Research and Regenerative Medicine (2018) 1: 157–160
https://doi.org/10.1007/978-3-319-77482-4

Index

Life Sci

Printed by Printforce, the Netherlands